LEAN MASTERY COLLECTION

8 BOOKS IN 1

Lean Six Sigma

Lean Startup

Lean Enterprise

Lean Analytics

Agile Project Management

Kanban

Scrum

Kaizen

JEFFREY RIES

© **Copyright 2018 by Jeffrey Ries. All rights reserved.**

This document is geared towards providing exact and reliable information regarding topic and issue covered. The publication is sold with the idea that the publisher is not required to render accounting, officially permitted, or otherwise, qualified services. If advice is necessary, legal or professional, a practiced individual in the profession should be ordered.

From a Declaration of Principles which was accepted and approved equally by a Committee of the American Bar Association and a Committee of Publishers and Associations.

In no way is it legal to reproduce, duplicate, or transmit any part of this document in either electronic means or in printed format. Recording of this publication is strictly prohibited and any storage of this document is not allowed unless with written permission from the publisher. All rights reserved.

The information provided herein is stated to be truthful and consistent, in that any liability, in terms of inattention or otherwise, by any usage or abuse of any policies, processes, or directions contained within is the solitary and utter responsibility of the recipient reader. Under no circumstances will any legal responsibility or blame be held against the publisher for any reparation, damages, or monetary loss due to the information herein, either directly or indirectly.

Respective authors own all copyrights not held by the publisher. The information herein is offered for informational purposes solely, and is universal as so. The presentation of the information is without contract or any type of guarantee assurance.

The trademarks that are used are without any consent, and the publication of the trademark is without permission or backing by the trademark owner. All trademarks and brands within this book are for clarifying purposes only and are the owned by the owners themselves, not affiliated with this document.

BOOKS INCLUDED:

Lean Six Sigma

Lean Startup

Lean Enterprise

Lean Analytics

Agile Project Management

Kanban

Scrum

Kaizen

Table of Contents

Lean Six Sigma

Introduction 13
Chapter 1: What is Six Sigma? 14
Chapter 2: The Different Levels of Implementing Six Sigma 19
Chapter 3: Why is Six Sigma Used? 21
Chapter 4: Tools to Use with Six Sigma 24
Chapter 5: Steps to Follow in the Six Sigma Methodology 32
Chapter 6: Scope the Perfect Project 36
Chapter 7: Transform Your Problem 39
Chapter 8: Knowing Your Goals and Your Needs 41
Chapter 9: Determine Who is Responsible for Each Project Part 43
Chapter 10: Picking Out the Solution, Implementing, and Following Up 45
Chapter 11: Common Issues with Implementing Six Sigma 48
Chapter 12: How to Get Six Sigma Certification 52
Chapter 13: Tips to Help with Six Sigma 56
Conclusion..61

Lean Startup

Introduction 63
Chapter 1: Lean Startup Options 67
Chapter 2: Create a Useful Lean Startup Experiment 76
Chapter 3: Growing a Startup 81
Chapter 4: Six Sigma Basics 90
Chapter 5: Implementing Six Sigma 96
Chapter 6: Additional Strategies 102
Conclusion 109

Lean Enterprise

Introduction 111
Chapter 1: Why Lean Matters to Your Enterprise 113
Chapter 2: Creating a Lean System 120
Chapter 3: Setting Lean Goals 125
Chapter 4: Simplifying Lean as Much as Possible 130
Chapter 5: Lean and Production 137
Chapter 6: Run a Lean Office 141
Chapter 7: Kanban 144
Chapter 8: 5s 147
Chapter 9: Six Sigma 152
Conclusion 157

Lean Analytics

Introduction 159
Chapter 1: What is Lean Analytics? 160
Chapter 2: The Lean Analytic Stages Each Company Needs to
    Follow.. 167
Chapter 3: The Lean Analytics Cycle 169
Chapter 4: False Metrics vs. Meaningful Metrics 173
Chapter 5: Recognizing and Choosing a Good Metric 177
Chapter 6: Simple Analytical Tests to Use 181
Chapter 7: Step 1 of the Lean Analytical Process:
    Understanding Your Project Type 187
Chapter 8: Step 2: Determine Your Current State 199
Chapter 9: Step 3: Pinpoint the Most Pressing Metric 203
Chapter 10: Tips to Make Lean Analytics More Successful for
    You 206
Conclusion 209

Agile Project Management

Introduction 211
Chapter 1: What Do We Mean by Agile Project Management 213
Chapter 2: How to Implement Agile Project Management 219
Chapter 3: Agile versus Waterfall Model 230
Chapter 4: Learning more about Scrum and Agile Principle 237
Chapter 5: Turning Your Organization Agile 244
Chapter: 6 Principles of Agile Plus Agile Manifesto 249
Chapter 7: Techniques of Agile Software Development 253
Chapter 8: Challenges of Implementing Agile 256
Chapter 9: Agile Methodology 260
Chapter 10: Keys to Successful Implementation of Agile 264
Conclusion.. 273

Kanban

Introduction 275
Chapter 1: The Current Status of Kanban 276
Chapter 2: How to Utilize the Kanban Process in a Non-Manufacturing Setting 280
Chapter 3: Applying Kanban to Lean Manufacturing 287
Chapter 4: Applying a Kanban Process to Software Development 293
Chapter 5: How Kanban Reduces Risk and Creates Improved Software 299
Chapter 6: Applying A Kanban Process to Workflow in Your Company 304
Chapter 7: Implement A Kanban System Effectively 309
Chapter 8: Implement Kanban Digital Boards for Production ... 316
Chapter 9: Development Tips for Your Kanban Digital Boards 321
Chapter 10: The Difference Between Kanban and PAR 326
Conclusion 331

Scrum

Introduction 334
Chapter 1: Basics of Scrum... 335
Chapter 2: The Sprint...341
Chapter 3: Looking Back on a Sprint and Planning for the
    Future 347
Chapter 4: Artifacts of Scrum 350
Chapter 5: Scrum Master as Servant Leader 354
Chapter 6: Making the Scrum Transition 361
Chapter 7: Tips for Success 364
Chapter 8: Stories from the Trenches..371
Conclusion... 375

Kaizen

Introduction 377
Chapter One: Kaizen and Teamwork.. 383
Chapter Two: Implementing Kaizen in a Startup 386
Chapter Three: The Five S's of Kaizen 398
Chapter Four: A Step-by-Step Kaizen Guide for Startups and
    Small Businesses 403
Chapter Five: Idea-Sharing and Kaizen Boards..........................414
Conclusion... 420

Lean Six Sigma

A Beginner's Step-By-Step Guide to Implementing Six Sigma Methodology to an Enterprise and Manufacturing Process

Introduction

Congratulations on getting a copy of Lean Six Sigma: A Beginner's Step-By-Step Guide to Implementing Six Sigma Methodology to an Enterprise and Manufacturing Process and thank you for doing so.

The following chapters will discuss everything that you need to know to get started with Six Sigma. Six Sigma is a methodology that is going to change the way that you do business. It strives to help you reach near perfection in the products that you sell, the customer service that you provide, and the lack of waste that you achieve. Moreover, it can work for all types of industries and businesses.

This guidebook will provide you with the tools you need to work with Six Sigma and see an improvement in your business. While other companies may waste hundreds of thousands of dollars on inefficient methods and faulty products, you can use the Six Sigma method to help improve your customer service, increase your productivity, and make your company more efficient. When Six Sigma is implemented properly, you can reach near perfection in all your company processes. This guidebook will show you how this is possible!

There are plenty of books on this subject on the market, thanks again for choosing this one! Every effort was made to ensure it is full of as much useful information as possible, please enjoy!

Chapter 1: What is Six Sigma?

If you have been running your business for some time, you have probably heard about Six Sigma at some point. It is known as a quality improvement method that is used often to help you find defects in the business model so that you can reduce them and get the business to be more effective than before.

Six Sigma is one of the most effective methods currently available to help improve the performance of any organization. It is able to do this by minimizing the defects in a business's products or service. With this method, all the errors committed have a cost associated in the form of losing customers, replacing a part, waste of material or time, redoing a task, or missing efficiency. In the end, this could end up costing the business. Six Sigma works to reduce these losses in order to help a business grow.

This methodology of Six Sigma was endorsed by Motorola in the 1980s. The company, at that time, was trying to find a way that they could measure their defects at a granular level compared to the previous methods, and their hope was to reduce these defects in order to provide a better product to their customers.

What they ended up with was a huge increase in the quality levels of several of their products, and the company received the first Malcolm Baldrige National Quality Award. It did not take long until Motorola shared their Six Sigma method and soon there were many other companies who were reaping the rewards as well. By 2003, it was estimated that the combined savings of all companies using the Six Sigma method were more than $100 billion.

What is the Sigma Scale?

The first thing that we are going to look at is known as the Sigma scale. This is a universal measure of the performance of any type of organization or business. Sigma is known as the statistical term to represent a standard deviation or the measure of a variation in a dataset. Higher scores of this indicate better performance, or it can mean the results are more precise.

In other words, if the output is defective 60 percent of the time, it means that the performance of One Sigma is compliant. However, if the output is defective 31 percent of the time, it means that it is demonstrating what is known as Two Sigma compliance. An example of how this works for the different Sigma's includes the following table:

\multicolumn{3}{c}{The Sigma Scale}
Sigma	Percent Defective	Defects out of a Million
1	69%	691,462
2	31%	308,538
3	6.70%	66,807
4	.062%	6,210
5	0.02%	233
6	0.00%	3.4
7	0.00%	0.019

As shown above, Six Sigma implies almost perfect output because there are only 3.4 defects per million opportunities.

DPMO

The term defect is used a lot when it comes to Six Sigma. The goal of the company is to reduce how many defects occur so they can reduce waste, provide a better product for their company, and make more money. However, what does this defect mean?

The "defect" is going to be explained as the nonconformities that are showing up in an output that falls lower than what the customers find as satisfactory. The number of DPMO, or defects present per million opportunities, is going to be used to figure out which part of the Sigma scale that process corresponds to. Most organizations in the world would fall somewhere between Three and Four Sigma. This may not seem so bad, but it really implies that they could be losing up to a quarter of their total revenue simply because there are some defects in the organization.

The Six Sigma methodology can help these businesses out. It can move them up to a new level of Sigma, which can reduce all that waste and those defects, effectively helping them to earn more profits.

Applying Six Sigma

While there are many different methodologies that can come with Six Sigma and can help the business to reduce its defects, the two basic ones that are good to start with include DFSS and DMAIC. Let us look at each of them below to understand how they work a little better.

DMAIC

The first one is DMAIC. In order to help modify a process that is already in existence and change it so that it can be more compliant with the Six Sigma methodology, therefore, more efficient, you would work with DMAIC. This stands for:

- Define: This is where an organization needs to define the goals for process improvements so that they are in coherence with the strategies of the business and with the demands of the customers. You can't get far in your process without defining what goals you want to reach and which processes must be improved to reach these goals.
- Measure: This is the current performance of the systems in place for the business. It will also take some time to gather data that is relevant and can be used in the future. Measuring the data you receive and the results that you are looking for can be important, and you must make sure you are relying on the right tools to do it.
- Analyze: This is where you will analyze the current setting and then observe how the relationship between the performance and the key parameters work. Lean analytics can be a good tool to use to analyze the situation and make sure your improvements are actually working. If changes need to be made, your analysis will showcase when this should happen.
- Improve: From the other steps, you will be able to find ways to improve the process. This helps to optimize the process to earn the business more money.
- Control: Here you will control the parameters before they are able to affect the outcome.

DFSS

Now, the business can also choose to completely start a new process from scratch and make it work with the Six Sigma methodology. This would be done with DFSS. When a new process is started in this manner, Design for Six Sigma methodology, or DFSS, is going to be used. Some see it just as an offshoot of Six Sigma and others are going to see it as its own methodology. Either way, the DFSS requires the IDOV approach, which stands for:

- Identify: Here you are going to identify and then define the process goals. These need to be consistent with both the standards of your industry as well as the demands of the customer.
- Design: This includes taking into consideration all of the possible solutions, and then selecting the one that is optimal.
- Optimize: This is when you will optimize the performance of the application. You can use different methods to do this such as statistical modeling and advanced simulations.
- Validate: This is when you will verify the solution that you chose and see if it works.

There are times when the DMAIC that we talked about before can turn into a DFSS. This happens when you look at the processes you are using and find they are completely irrelevant or just are not working for you. You may decide to redesign them to get the results that you want.

For some people, they want to be able to use the system that they already have in place to get started with Six Sigma. Others find that how they currently do things is such a mess that it is just easier to start over and try something new. Your business will have to look at its current processes and determine which method will make the best sense for you.

Chapter 2: The Different Levels of Implementing Six Sigma

There are several different levels when it comes to implementing Six Sigma. The role that a person plays in this is going to depend on how much training they have with Six Sigma, what position they hold in the company, and so much more.

Implementing Six Sigma

Six Sigma is designed to have professional quality management roles for everyone who is a part of the team. They have also adopted a ranking system that is similar to what you find in martial arts, in order to tell who does what in the business when working with Six Sigma. Some of the different rankings for this method include:

- Executive Leadership: This is going to be anyone who is in the top management level of a business and can include the CEO. These visionaries are going to authorize the others on the team to provide them with the needed resources to improve a process. You must make sure that your leadership includes some of those in the upper levels of your company. Without this support, it can be impossible to get Six Sigma and its processes to work properly.
- Champions or the Quality Leaders: These individuals are going to be in charge of integrating and implementing the methodology through the whole organization. They are the ones who will be chosen from the upper management and they can be mentors to the Black Belts.
- Master Black Belts: These are individuals identified by the Champions and they are expert coaches with Six

Sigma. They are going to be guides to the Black Belts as well as the Green Belts and they will assist the Champions in implementing Six Sigma throughout the organization.
- Black Belts: These individuals will operate under the guidance of the Master Black Belts. Their main task is to supply the ideas of Six Sigma to the different projects specified. They can also be in charge of executing the Six Sigma project. They will need direction from the Master Black Belts, but they do have some leadership qualities as well to perform their jobs.
- Green Belts: These individuals will work on the implementation of Six Sigma in the business. They can also work with other projects in the business. The belts above this one will devote all their energies to Six Sigma, while these individuals will be able to work on other projects that the business needs, but can also work with Six Sigma a little bit.
- Yellow Belts: These employees have some training with the techniques of Six Sigma, but they do not really have the opportunity to apply their knowledge to a project that is using or relying on Six Sigma.

Each of these levels is going to be important to implementing Six Sigma for a business. This is not something with which just one or two people can do and be effective. This large methodology can help reduce waste, create better products, and help a company to meet and exceed the expectations of their customers. However, it is going to take some work from top management, as well as from others who are well versed in Six Sigma, in order to see the great results. Everyone can play a role in Six Sigma, which is why it is so important for a company to implement it throughout the whole business and ensure everyone is a part of it.

Chapter 3: Why is Six Sigma Used?

Now that we have looked at Six Sigma and what it is about, you may have some more questions. A common one is "Why is Six Sigma used in the first place?" Is it that good at getting rid of defects in a business? Can it really help to reduce waste as much as it claims? Or is it too much work to implement? There are many reasons why a business would choose to implement Six Sigma, and some of the main ones include:

- The quality of the product that the company provides to customers will improve by quite a bit. The same can be said when talking about the productivity of the company.
- The number of possible defects that happen out of a million opportunities will reduce quite a bit. This means that the quality of a product is going to improve way more than a company can do on its own. The fewer defects in a product, the happier the customer is.
- The result of any process that is defined here is going to be based on data that has been collected, rather than just on some assumptions that management tries to make.
- The amount of profit that the company is able to make is going to increase rapidly. This means that the company now has growth in the terms of more profits and more opportunities.
- It calls on the business to come up with a more correlated and integrated approach to help them solve problems that are already there.
- Some of the other conventional cost-cutting methods are not going to be the best for some companies. Six Sigma will prefer to remove costs that end up not giving

value to their customers.
- The net production costs that you will incur when you manufacture a new product will be reduced as well.
- By being able to reduce how many defects they are dealing with effectively, a company may be able to raise their expectations in the future.
- Six Sigma can help to meet the needs and the expectations of more of their customers. This is because the company can use this method in order to provide them with the same product they love, but with higher expectations that it will be good, last a long time, and do what it promises.
- The internal understanding inside of a business across the various departments is going to increase. This means that the employees of a business are going to be more aware of the techniques and strategies that each part uses in order to solve a problem. This helps them all to learn and do better with their own work.
- There seems to be more job satisfaction for employees when Six Sigma is used. Because of this job satisfaction, the amount of internal communication that occurs inside of an organization is going to increase as well.
- The time spent on production while manufacturing a new product is going to decrease. This allows the business to deliver the service faster than ever.
- The market value of that company will increase due to all of the other factors.

As you can see, there are tons of great benefits that come with Six Sigma; it is a tool that many businesses are using so that they can provide a better product to their customers and make more money in the long-term.

The other side of Six Sigma

Even though there are many reasons to love Six Sigma, some people do not rank this process very high. Six Sigma sets up a standard of 3.4 defects per a million opportunities. For many businesses, this sounds fantastic because it can help them to provide better products and services to their customers.

However, this type of quality standard is not always the optimal one to go with depending on your process. For example, if you are working with a process that concerns the well-being of a person, you may want to go with a quality standard that is even higher.

In addition, there are also some other types of businesses and operations that are going to work well if they have a lower level of Sigma. Many find that there really isn't a clear justification as to why a business would choose to go with the number six from the table, and they may do better if they go above or below this level. Because of this capriciousness, there are some quality managers who are not as fond of using Six Sigma and they will choose a different method to get their defects taken care of.

Chapter 4: Tools to Use with Six Sigma

There are many tools that you can use in order to make Six Sigma work for you. These tools are there to ensure that you are providing good quality management to your business and some of the tools are so successful that they can be used outside of a Six Sigma application as well. Some of the main methods that can be used include:

5 Why's

The 5 Why's is a technique that is there to explore the cause and effect relationship of a problem. The goal of this technique is to find out the root cause of a problem by repeating the question "Why?" Each answer is going to form the basis of the following question. The 5 in the name derives from the idea that it takes about five iterations in order to resolve the problem, but depending on your particular issue, you may need to use more.

Not all problems though are going to have one root cause. If you would like to figure out more than one root cause, this method is going to be repeated by asking a different sequence of this question each time that you use it.

In addition, the method is not going to provide any hard rules about what lines you should explore with the questions, or how long you need to continue your search to make sure you find the root cause. Thus, even if you follow this method closely, it may not give you the outcome that you want.

An example of the 5 Why's includes the following:
My vehicle is not starting:

1. Why? The battery is not working.
2. Why? Because the alternator is not functioning
3. Why? The belt on the alternator has broken off.
4. Why? The belt should have been replaced a long time ago, but was not.
5. Why? The vehicle owner did not follow the required maintenance schedule for the vehicle.

This helps to show why there was an issue with the vehicle, and you can easily choose to take it further into some more why's until you find the solution that you are looking for.

Axiomatic design

The axiomatic design is a systems design methodology that is going to analyze the transformation of the needs of the customer into design parameters, functional requirements, and process variables. The method is going to get its name because it is going to use the design principles that govern the analysis and decision-making process. The two types of axioms that are used with this process include:

Axiom 1: This is the independence axiom. It is going to help you to maintain the independence of your functional requirements.
Axiom 2: This is the information axiom. This is going to help you to minimize the informative content of the design.

Cost-benefit analysis

Cost-benefit analysis, or CBA, is an approach that is meant to estimate the strengths or weaknesses of varies alternatives. It can be used with project investments, processes, activities, and even transactions. It can be used to determine, out of several solutions, which options will provide the best approach to a business in order to achieve benefits while still saving the company money.

To keep it simple, the CBA method is going to come with two main purposes. These purposes are:

- To determine if a decision or an investment for a business is sound. This means that the benefits will outweigh the cost. You also want to look at how much this is. If the benefits do not outweigh the costs much, then it is probably not the best option to go with.
- To help provide a good way to compare projects. This can involve comparing the total amount that you expect each option to cost against the benefits you expect to get.

The benefits, as well as the costs, are going to be shown in monetary terms, which makes it work well for Six Sigma. Moreover, they can be adjusted in the formula for the time value of money. This ensures that all flows of costs and those from benefits over time are expressed with a common basis.

The simple steps that you will follow when you are working on a cost-benefit analysis include:

- You first have to define the goals and the objectives of the project or the activity.
- You can list the alternative programs or projects that you may be able to use.
- List the stakeholders
- You then select the measurements you want to use in order to measure all of the elements when it comes to benefits and costs.

- You can also work on predicting the outcome of the benefits and the cost of each alternative over a period of your choosing.
- You can then convert all of the benefits and costs into a common currency to help them compare better.
- Make sure to apply any discount rates
- Next, you can calculate the net present value of all project options.
- Perform a sensitivity analysis: This is going to be the study of how the uncertainty of the output from a mathematical system can be shared to different sources of uncertainty in its inputs.
- After you have all this information, you can then pick out the option that is the best.

Root cause analysis

A root cause analysis, or RCA, is going to be a method to help with solving problems and it focuses on finding the root causes of the problem. A factor will be considered the root cause if you can remove it and the problem does not recur. Essentially, there are going to be four principles that come with this type of method including:

- It is going to define and describe properly the problem or event.
- Establish a timeline from the normal situation until the final failure or crisis occurs
- Distinguishes between the casual factor or the root causes
- Once it is implemented, and the execution is constant, the RCA is transformed into a method of problem prediction.

The main use of the RCA is to identify and then correct the root causes of an event, rather than just trying to address a symptomatic result. An example of this is when some students receive a bad grade on a test. After a quick investigation, it was found that those who took the test at the end of the day ended up with the lower scores.

More investigation found that later in the day, these students had less ability to stay focused. In addition, this lack of focus is from them being hungry. So, after looking at the root cause and finding it was hunger, it was fixed by moving the testing time to right after lunch.

Notice that the root causes are often going to come in at many levels and that the level for the root is only going to be where the current investigator leaves it. Nevertheless, this is a good way to figure out why one particular process in the business is not working the way that you want and then finding the best solution to fix it.

SIPOC analysis

If you are talking about process improvement, a SIPOC is there to be a tool that can summarize the inputs and then the outputs of at least one process and then shows it in table form. This acronym stands for suppliers, inputs, process, outputs, and customers and these will be used to form the columns on your table.

Sometimes the acronym is going to be turned around in order to put customers first, but either way, it is going to be used in the same way. SIPOC is presented at the beginning of a process improvement efforts or it can be used during what is known as the define phase of the DMAIC process. There are three typical uses of this depending on who is going to use it including:

- To help those who are not familiar with a particular process a high-level overview.
- To help those who had some familiarity with the process, but may be out of date with the changes in the process or those who haven't used it in a long time.
- To help those who are trying to define a new process.

There are also some aspects that come with this method that are not always apparent. These include:

- The customers and the suppliers are sometimes external or internal to the organization that is trying to perform the process.
- Outputs and inputs can include things such as information, services, and materials.
- The focus of this method is to capture the set of inputs as well as outputs, rather than worrying about all the individual steps that are in the process.

Value stream mapping

When it comes to value streaming mapping, we are talking about a method that is there to analyze the current state of a business and then designing a new state to use in the future. It is meant to take a service or product that a company offers from its very beginnings all the way through to when it reaches the customers. The hope is that the process is used to help reduce lean wastes, especially when compared to the process that the business is using right now.

The value stream is going to learn how to focus on any areas of a business that helps to add in value to the service or product. The purpose of this is to learn where the waste is in the business and then remove or at least reduce it. This can increase the efficiency of the business and can even increase productivity.

The main part of this process is to work on identifying waste in the business. Some of the most common types of waste include:

- Faster than necessary pace: This is when the company tries to produce too much of their product that it can damage the flow of production, the quality of the product, and the productivity of the workers.
- Waiting: This is a time when the goods are not being worked on or transported.
- Conveyance: This process is used to move the products around. It can look at things like excessive movement and double handling.
- Excess stock: This is when there is an overabundance of inventory. This can add on storage costs and can make it more difficult to identify problems.
- Unnecessary motion: This waste means employees are using too much energy to pick up and move items.
- Correction of mistakes: The cost that the business will have when they try to correct a defect.

This process is used often in lean environments to help look at and design flows for the system level. This is often something that is associated with manufacturing, but it can be used in many other industries including healthcare, product development, and even software development.

Business Process Mapping

The idea of business process mapping is going to be all activities that are involved when you try to define what a business does, who is the person or persons responsible, and at what standard a process in the business needs to be completed. It can also determine how the success of the process in the business can be measured.

Business process mapping is there to help a business become more effective. A clear business process map will allow even outside firms, such as consultants, to come in and look to see where improvements can be made, such as what can happen with Six Sigma, to help the business.

This mapping is going to take a specific objective of a business and they can measure and compare it to the objectives of the company. This makes sure that all processes that are done can align with what the company holds as its capabilities and values.

A good way to do business process mapping is with a flow chart. This can help you to see how the business does a certain process and can even include who is responsible for each part if that is important.

These are just a few of the options that you can choose from when it comes to working with Six Sigma. All of the options above can help you to make informed decisions while finding the process that is causing your business the most trouble at the time. Pick one of these options that go along with your biggest issue and find out how you can make smart decisions that will turn your business into something even better.

Chapter 5: Steps to Follow in the Six Sigma Methodology

While some companies decide to start from the very beginning with a new process in the hopes of getting fewer defects and providing better service for their customers, this is not always something that needs to be done. It is likely that your business already has a process in place. Moreover, with a few simple tweaks, you will be able to make the right changes that can make it Six Sigma compliant. This is why most companies will use the DMAIC method to help them with Six Sigma.

The steps that are used in the Six Sigma methodology are there to help you adopt a way of doing things that is smarter so you minimize the occurrence of defects. It has an emphasis on getting rid of waste and doing things right the first time. This provides you with a better customer experience and can help you save a lot of time and money fixing the defects later on.

While this may seem logical, you may be asking how you would do this in a way to ensure you did it right the first time. The steps in this methodology are going to ask you to adopt a simple five-stage process, which includes:

Step 1: Define

For this first step, you are going to look at the process or the data and find the area or the process that needs improvement. This is the nature of your problem. During this step, you will also form a team and help them train in the Six Sigma method so that they can work with you to improve the current process. You must make sure that you pick out a team that is motivated and believes that Six Sigma is important, otherwise things could get messed up a bit.

The next thing to do is to identify the customers or the people who would be the most impacted by this project. You can also document the critical requirements for these customers. Then you can create a team charter that is going to detail things like the business case, the project scope, and the statement about the problem. This is going to help you finish up the define step.

There are some different tools that you can use during this phase to make it a bit easier. Project Charter, SIPOC, and stakeholder analysis are all good diagrams that can help you to depict the different elements of your new project visually.

Step 2: Measure

You will find that the measuring step is going to take a bit more time compared to the defining stage. It is during this particular stage that you are going to define what parameters you will use in order to measure how to see whether performance has improved. You will also define the baseline performance as well as the extent to which the process can be improved.

You may spend some time in this phase looking at and identifying the key defects in the process. Then when you define the key measures to improve, the data is going to be collected so that you can analyze the differences between the desired performance that you want and the current performance that you already have. You should also take some time to establish the process variations during this phase.

Step 3: Analyze

The third step that you are going to work on is to analyze. During this particular phase, the data that you collected in the past phase will be used in order to analyze what the gap is between the current and the desired performance. You can then do a root-cause analysis to help you determine what can be causing the gap between your current performance and the goals that you want to reach. This is often going to be calculated using financial terms so you can see the issue in dollars.

Step 4: Improve

Now that you have had some time to define the problem, measure the gap that is there with performance, and then analyze the reason for this gap, it is time to move over to the fourth step. During this one, we are going to take the steps that are necessary in order to improve the issues that are there.

During this particular phase, you are going to devise a set of solutions that you could use. Sometimes there may be only a few options to work with, and other times there may be many different options from which to choose. From there, you can pick out the best possible solution based on the Six Sigma method and the options you have decided on.

The main outcome that you want to get out of this phase is to design a performance improvement plan. This plan is supposed to work in order to provide you with a measured difference in your existing process so that you can really see your defects go down.

Step 5: Control

The final phase that you can work with in the Six Sigma methodology is the control phase. Here, you will come up with the project management plans as well as any procedures that need to be followed in order to sustain the new process you created.

Many companies forget this step, but it is very important. How is the Six Sigma method going to work if no one follows the plan that you come up with along the way? During this phase, you will need to document the process that you revised, devise and then deploy your response plan, and then transfer this information about the new process over to the management and others who need to use it.

When all of these parts come together, you will find that the Six Sigma method can work really well. It helps you to figure out where the problem areas of your company are and can provide you with data in order to come up with a plan to fix them. It can be time-consuming and you need to make sure that everyone in your company is on the same page when it comes to using Six Sigma so you can get the best results out of it for your company.

Chapter 6: Scope the Perfect Project

Now that we have some of the basics down about working with Six Sigma, it is time to help you get started with a project using Six Sigma. The best way to learn how to use this process is to actually get a project and get to work. You can read about it all that you want, but it is hard to understand until you get a project in hand and can get started on it. This chapter is going to look at the first steps that you can take when working on a Six Sigma project: scoping the perfect project.

Scope the project

So, a Six Sigma project is going to start out as a practical problem that is affecting a business adversely. Then it is going to end as a practical solution that can help to improve how a business is able to perform. If you can find a project where a business needs some help with their current processes, then you may have a good option for Six Sigma.

The focus of your project with Six Sigma is going to be to solve a problem that could be hurting some key elements of performance for a company. Some examples of these could include:

- Process capability
- Costs
- Customer or employee satisfaction
- Organizational viability
- Revenue potential
- Cycle time
- Output capacity

You will want to get started on the project by stating out what the problems with performance are. Make sure you use terms that are quantifiable and that are going to define expectations. These also need to relate to the levels of timing and performance that you desire in the end.

As you are going through and defining the project that you want to use, you should also pay attention to some important issues. For example, some are going to be especially good for warranting a Six Sigma level of effort. Some problems that you can consider include:

- Issues that are going to have an impact on Earnings Before Income Tax, or EBIT, or Net Profit Before Income Tax, or NPBIT. You can also look at those that have a significant strategic value.
- Issues that are going to give you results that seem to exceed the effort that is required in order to see improvement by quite a bit.
- Not easily or quickly solvable if you use some of the other methods that you have seen in the past.
- Issues that are going to improve the Key Performance Indicator by more than 70 percent over the existing levels of performance where they are now.

There can also be a type of flow that comes to your project as well. You will want to flow in the order below to help you work on a Six Sigma project in the current manner:

- Practical problem: This is a chronic or otherwise systemic problem that is affecting the success of your process.
- Six Sigma Project: This is going to be an effort that is well defined and that states your problem out in terms

that are quantifiable and which have known expectations.
- Statistical problem: This is a data-oriented problem that will use data and facts to help figure it out.
- Statistical solution: This is again a solution driven by data and has known risk and confidence levels. This is in comparison to "I think" solutions that may have been used in the past:
- Control plan: This is a method that you can use that will assure the long-term sustainability of your solution for the problem. You do not want to come up with a solution that seems to work today, but then does not work a few months or so down the line. You want to go with a solution that can work for in the long-term and still provide you with some great results.
- Practical solution: This type of solution is not seen as irrational, expensive, or complex. The best part is that it can be implemented without a lot of problems or a wait time.
- Results: These are the tangible results that you get. You will be able to measure it out financially or in other ways that show how it is benefiting the company.

Getting started on a new project can sometimes be the hardest part. You want to make sure that you are working on a project that can help the business out, and that will let you use the Six Sigma methodology to get it done.

Chapter 7: Transform Your Problem

The first step that we talked about was to scope out the project that you want to work on in a business; it is now time to move on to the second step. After you have been able to take the particular problem or process that you want to fix in your company and framed it to be a potential project with Six Sigma, it is time for the problem to go through a change. It is going to transform from a business problem that is practical over to what is known as a statistical problem.

The reason that it changes between these two types of problems is so that you are able to identify a statistical solution. This is easier to understand in some cases, and you will then be able to move it back to a practical solution after you receive the data and the information.

This is why when you define the project, you need to make sure that you are stating it out using some statistical language. This ensures that you and others on the team are going to use data, and only data, to help solve it.

Data is not going to lie to you unless you knowingly choose to read it or use it in the wrong way. You will get real results from the data, and you can make better decisions from it. Often changing the problem into some form of data is going to make a big difference in how big the problem seems to you. You may see that customers are not coming back to you, but when you can see how many customers you have lost, or how much this translates to in dollar amount, you may see that it is actually quite a bit of the problem.

Data also helps you to use the facts at hand. Too many times in business it is easy to use your intuition, some best guesses, and even your gut feelings in order to address the problem. This is a big issue when you want to improve the efficiency of your business and get results. This is what Six Sigma can do to help you succeed.

You will quickly realize that, in business, you are not able to solve real-life business problems simply by throwing your time and your money at them. This seems to be the way that some businesses are choosing to handle things, though, and it is not going that well in terms of customer satisfaction and profits for them.

What you need instead is some practical solutions. A project that uses Six Sigma can provide you with a good solution that is not complex, will not be too hard for you to implement into your current business model, and will not require extensive resources in order to get it to provide you with the improvements that your business needs.

Transforming your problem is one of the best steps that you can do. It will give you some data that will help you to see the full extent of the project and will ensure that you are actually handling the project. You will then be able to take that data and turn it back into a practical solution later on, but for now, just find ways to turn the problem that we discussed in the first chapter and turn it into a statistical solution.

Chapter 8: Knowing Your Goals and Your Needs

In order to get the maximum benefits from your Six Sigma projects, you need to be aware of things such as the strategic goals, objectives, and needs of the business. These always need to be in your mind when you are trying to figure out which problems should be solved with this process.

You will start out this step by finding out which areas of the business must be improved in order to meet the specific goals of the business. Of course, if you do not know these, you will need to meet with the top management of the company in order to figure out what they are. Otherwise, you will start trying to solve problems that lead the business away from its goals.

This approach is going to lead you to determine which problems specifically you need in order to improve the performance of a company. Then you can move on to determine the statistical solution to the problem before implementing the solution and realizing the benefits from the work.

From here, you may be confused as to where you should start. A good place to start is with an assessment of the higher-level needs of the organization. You can also use any knowledge that you can get from the voice of the people and voice of the business. With the VOC, you are looking at all the expectations and needs that the customers already have for your products. On the other hand, the VOB is when you look at the expectations or the needs of a business.
The basic idea that you need to keep in mind is that you should

do an assessment of the VOB and the VOC in order to see where there are any gaps. These gaps are going to be areas where the expectations of the customer and the expectations of the business do not line up well together.

Zeroing in on the problem areas is going to be key to helping you out here. You also need to have all the information organized and present to help you with this. To make things easier, you can always look for some themes in the information, such as:

- If there are any issues with invoicing or with accounts receivable
- If there are any issues with capacity constraints
- Look through any complaints from the customers and see if there are some common threads or themes with those that the company can address
- The responsiveness or the cycle time of the business and if it aligns with what the customer expects
- If there are excessive levels of inventory for the business
- If there are any services that are defective or ineffective for the customer
- If the yield or subsequent rework or scrap that you need to work on
- How many times a product is returned and the costs for warranty and how these add up for the company and could be costing them more money in the long run

All of these can be issues for a company, but they often have one or two that seem to cause the biggest waste and loss of profits. Being able to know the goals and the needs of a company, and then zeroing in on the ones that need to be improved the most will help you to make Six Sigma work the best for you.

Chapter 9: Determine Who is Responsible for Each Project Part

In addition to working on the steps that we have talked about in the other chapters, you need to be able to determine who is going to be responsible for each part of the project. Problems that may start out in the functional area can transform from line managers on through to the owner.

Project deliverables, accountabilities, and responsibilities should be divided between managers and the different Belts (the yellow, green, black, and master black that we talked about before), who are going to need to perform a variety of activities for problem-solving. Managers, including the process owner, will be in charge of figuring out the focus and the priority of this process.

On the other hand, the non-management personnel are going to be the ones who are responsible for implementing the solution that was determined before and then seeing the benefits.

All of these parts need to come together. The managers and the nonmanagers need to all work together through the Six Sigma methodology in order to see success. These relationships can help ensure that the deliverables from Six Sigma are not going to end up falling through the cracks.

As you can imagine, the whole methodology of Six Sigma needs to be a team effort. Even when you are just at the beginning and working on the Define phase, the phase where managers will identify the project to work on, the Belts are still there to assist.

Of course, those who are the Belts are only going to have about 20 percent responsibility when it comes to defining and then managing the improvement, while their managers will be closer to 80 percent. Later, when you reach implementation, these percentages are going to reverse as the Belts do the work to make sure the solution is working to fix the problem. This is why both parts are so important and why they need to be involved in all the parts of the Six Sigma process.

The managers, including the owner, will be able to determine who is in charge of each different part when it comes to doing each part. They can assign based on the job that needs to be done, which belts they have present in the business and more. However, before the project is even started, it is important that everyone is on board and they understand what needs to be done. This ensures that Six Sigma can be done properly and that nothing slips through the cracks.

Chapter 10: Picking Out the Solution, Implementing, and Following Up

One way to ensure that you can implement the Six Sigma project is to start with a pilot project. This helps you to figure out what problems are present in the business and what you can work on for them. You can also determine who will carry out the work and get their input through it all. Remember that Six Sigma is only going to work well if everyone in the business is involved in the whole process.

From here, we need to not only look at the problems that are going on with the business and that need to be improved, but we also need to pick out a solution, implement that solution, and make sure that the proper follow up occurs so that things don't get messed up or go back to how they were before. Let us look at each of these parts and see how they work with the Six Sigma methodology.

Picking out a solution to your problem

At some point, you are going to need to come up with some solutions to the problem you are choosing to fix. How you do this is going to depend on the problem you are fixing, but often there are more than one available solution from which to choose. So, how are you going to choose the right one based off all the choices you have?

This is where you need to go back and look at some of the tools that come with Six Sigma and use them with your data. This way, you will easily be able to see what will be the best solution to make you more efficient, reduce waste, and help you to make more profits in the process. In the end, you should end up with just one solution to handle the problem.

Implementing the solution

Now that you have taken some time to look at all of the different solutions that are available and you have chosen the one that you would like to work with, it is time to implement the solution. This is the step where everyone really needs to be on board with each other. If not, things are not going to be implemented right and your Six Sigma methodology will not work the way that you want.

You and the others on your team can come up with the procedures that need to be followed in order to come up with a good implementation of your solution. This could be new training for the employees, or a new way that you do the manufacturing of a product. It could be many other things. Nevertheless, you have to look at the solution and find a way to implement it so everyone is on the same page.

Following up to see if it works

Hopefully, the solution that you decided to implement is going to work well for you. It will be a solution that you can implement easily and that your employees and customers will enjoy. It will take away some of the issues that you were having before and will ensure that you are able to be as efficient as possible with your business.

However, in order to make sure that this is working for your business, you need to have plenty of evaluation and follow up. This helps you to see that the solution is being implemented properly and that things are actually getting fixed.

As your project finishes up, an evaluation is important because it is going to detail what worked well and then what was causing you some problems. The workers who help with this are going to be a big source for you when it comes to evaluation parameters and criteria.

You will need to have some method in place to help you ensure that the new procedures are implemented the right way. Your managers should know how to do these evaluations and how to report them back, and even fix the behavior if needed.

The whole point of doing Six Sigma is to make sure that you get rid of the defects of your business and provide better service to your customers. If you come up with the solution and do all of the other parts of this process, but then you do not follow up and make sure that the solution you picked is actually being used, then you just wasted a lot of time and effort in the process.

Six Sigma is a process that is going to take a lot of time and effort on your part. It is not always the easiest thing in the world and it can cost money and time to make happen. However, if you are ready to provide a better experience for the customer, provide the best products possible, and reduce your overhead so you can make more profits, then Six Sigma is the option for you.

Chapter 11: Common Issues with Implementing Six Sigma

Many people in the business world have been talking about Six Sigma and how great it has been for them. Many of the organizations who choose to implement this method have found that it can improve their processes, services, and products. Having the ability to reduce their defects has helped these companies to increase their profitability, customer satisfaction, and productivity.

However, there are going to be some obstacles that can show up, and these will often stand in the way of being able to implement Six Sigma properly. You need to make sure that you recognize some of these hurdles and that you know how to address them if you want to be able to make Six Sigma work for you. Here are some of the common issues that sometimes come up when you are implementing Six Sigma and how you are able to fix them.

Lack of commitment from leadership

Six Sigma is not a methodology that you can read about once, hand off to the employees, and hope that it goes well. There needs to be a big effort among the whole company to make Six Sigma work, and getting commitment from everyone in management is one of the most important steps.

A true test to see if a company is truly committed to working with Six Sigma is going to come when the management decides which employees it is going to dedicate to the new Six Sigma project. It is always best to go with your top talent on a Six Sigma project, rather than just picking out whoever is available.

If you just choose to use some random people in the business who are available but do not have the right training, then you are already starting the project off on the wrong foot. Moreover, this can reduce how likely it is that the project will be a success. A successful Six Sigma project is going to require leaders who have the dedications to provide money, talent, time, and resources to this new endeavor.

You will find that taking your top performers and reassigning them from their current work so that they can work with the Six Sigma project is going to be the best bet. Sure, you may have to make some short-term sacrifices to make it work. However, in the end, you will end up getting the most benefits out of Six Sigma, which can greatly improve your business, when you choose to do this.

Not understanding how Six Sigma works

It is hard to get a methodology to work well if you do not have a complete understanding of how it works. Some organizations try to do it simply because they think they should. Some do it because they see that someone else was successful, and they want to be successful too. Others just rush into the project because they are so excited, but they do not have a firm grasp on what it requires to implement Six Sigma successfully.

The first thing to understand is that you should never implement the Six Sigma methodology just to keep up with the competition. You should also not waste your time on it if your only reason is to impress the shareholders. If you are only planning on using Six Sigma as a cosmetic change, or you implement it without giving it all the resources that it needs, then you are just inviting failure into the project from the start.

The best way to overcome this kind of obstacle is to commit to the process fully from the beginning. You can make sure that you employ and support the Six Sigma experts on your team to ensure that the new process starts working, rather than just bringing it on to use the terminology.

These experts, who may work for the company or not, are there to keep the project focused on the areas where they can make the biggest difference. Do not waste your time or the time of your team by using Six Sigma to help with simple changes that will not make that much of a difference. Six Sigma can do some amazing things for your business, but you have to be willing to take on the big tasks, the ones that really need to be changed, rather than the smaller options.

Poor execution

Even if you have some expert guidance to help with this process, there are times when the project is not going to go well because it was not executed properly. This poor execution will happen when the improvements are not aligning with the goals of the organization. It can be an issue when the project is reactively solving problems, rather than meeting strategic objectives. Alternatively, it can be when the quality improvement project focuses too much on the outputs of the project rather than focusing on the inputs.

When companies that use Six Sigma are able to understand that these methodologies are not there to act inside a vacuum but are there to work while aligning to the objectives and goals of the organization, they will find that it is easier to stay on target.

If you are working on Six Sigma and find that you are not getting the gains in productivity that you were hoping for or the savings financially, then it may be time to look for a reason. The reason is not going to be that the methodology of Six Sigma is ineffective. Instead, it is going to most likely come from a lack of effective leadership and the fact that the project was not managed in the way that it should've been.

When the leadership learns how to be committed and completely on board with the methodology of Six Sigma, they work to assign the top talent to their teams. They put the project through all the right steps, including the formal selection and the review process, and it provides the required resources to all people in the process. The odds of seeing success with Six Sigma are then going to increase quite a bit.

If Six Sigma is used in the proper manner, you are going to see some amazing results. You will be able to cut out wastes, help your customers out better, and make more profits in the long term. However, there are times when a business will take considerable resources to work on Six Sigma, and they will not see the results they had hoped for. Making sure that these top issues are not a part of our project can increase your chances of seeing success.

Chapter 12: How to Get Six Sigma Certification

Six Sigma is a project management methodology that is there to help increase profits, ensure product quality, boost morale, and reduce defects. Many companies use it in order to help them strive to be as close to perfect as possible. Although there is not really a governing body that will dictate the rules of Six Sigma, many organizations are going to offer certification in this methodology. By becoming certified with Six Sigma, you will find that you are someone to take seriously and that could provide more value to the company you are already with. Some of the things that you need to do in order to get Six Sigma certification includes:

Determining the management philosophy

1. Consider what the organization needs: What kind of management style is going to benefit your organization the most? Is it dealing with too much waste or overhead in the supply chain? Are there some issues with staying consistent to get things done? What is the overall culture in the business?
2. Decide how you would like to optimize the process: You may be someone who thinks that the best way to make sure quality is there is to ensure that all the processes are consistent with as few variations as possible. Others may want to opt for efficiency or producing a quality product without as much waste as overhead.
3. Determine which certification you want to go with. You can go with Lean Six Sigma or Six Sigma. The type of management philosophy you go with will determine the answer to this.

a. Six Sigma is going to define the waste as a variation in the process of the business. If you believe in a consistent process, then you should get this certification.
   b. Lean Six Sigma is going to be a combination of the Lean method and the Six Sigma method. It is going to define waste as anything that does not end up adding value to your product when it is done. If you would like to be more efficient, then this is the best option to choose.

Decide the level of certification

1. Determine your role in an organization. This can determine how high of a certification you should get. Are you someone who supports the manager or the manager? Does your work involve you just being able to use Six Sigma on a project?
2. Consider your goals in the future: Even if you are not a project manager right now, if that is your future goal, then you should consider this when picking out your certification.
3. Select the certification: There are four levels that you can choose including Yellow belt, Green belt, Black belt, and Master Black belt:
   a. Yellow Belts: These are those who have just a basic understanding of the process. They are more of a supporting role to those with the higher belts. There aren't many courses for this because it is so basic and most concentrate on being an expert in the field.
   b. Green Belts: These are the individuals who will work closely with the Black Belts and are mainly responsible for collecting data. These individuals will use Six Sigma, but they often have other

responsibilities outside this project, so they still just need a basic understanding.
  c. Black Belts: These are individuals who are project managers. The other individuals will report to them and they are going to be the ones who are going to dedicate a lot of time to the project.
  d. Master Black Belts: These will be the experts of the team. They will be the one that the team will turn to if there are any errors or if they need to make some corrections along the way.

Getting certified

1. Find a training program: It is likely that you will have to do some classroom instruction, so look and see if there are some near you to avoid travel. Always make sure that the program is accredited. No, there are no formal standards right now, but there are some accreditation organizations that can make sure you actually learn what you need.
2. Enroll in the program: You will attend the right classes and learn the material that you need to get the belt that you choose.
3. Take the written test: Once you are doing with the training, the next step is to do the written test. This will check to see if you have learned what you need about Six Sigma. These tests can take some time. The Yellow Belt can be two hours, the Green Belt about three hours, and the Black Belt about four hours.
4. Complete the projects: The final phase of being certified will be the process of completing a few projects using the Six Sigma methodology. This is like the "lab" to make sure that you are able to implement what you learn.

Benefits of being certified

Employees who are certified in Six Sigma can bring many benefits to a company. You already know how the process works and can be there to help the project go smoothly. It does not matter which belt you end up getting, they all are important to implementing a project and ensuring it is done.

If your current employer is looking at getting started with Six Sigma, it is definitely worth your time to get a certification. This can ensure that you are put on the project and can really help to further your career, especially if you see success. If your current employer is not getting started with Six Sigma, it is still worth your time. You may be able to use this later on with another company or whenever your current company decides to look as well.

That is all there is to it. The amount of time that you take to complete the certification is going to vary based on which belt you want to get and how the training centers work in your area. With some time and studying, you can learn how to make Six Sigma work for your business.

Chapter 13: Tips to Help with Six Sigma

Six Sigma is a methodology that can do many things for your business. It is there to help reduce some of the wastes that can come up in your business, helping you to provide better service and better products to your customers. However, there are times when you may not know exactly what to do to ensure you get the most out of Six Sigma. Some of the tips that you should follow when you first get started include:

- Make sure that leadership commitment is there. Make sure that the top management of the business is committed by getting them trained. This can talk about the responsibilities of the management as the Champions, tools used, and an introduction of Six Sigma. They need to be completely convinced about the benefits of Six Sigma. In addition, your steering committee should be formed to ensure:
    - The organizational goals of the business align with Six Sigma projects
    - The resources are all planned for and all roadblocks are out of the way
    - One person needs to be there to lead them all and they will be the Black Belt. Pick someone who would be top of your organization and can do the best job.
- Make sure your leaders are trained. They should be at least Six Sigma Champions. This two-day training session ensures that they are ready to help run the group.
- Elect the right person to train all the belts. There are tons of programs out there that promise to be the best, but most of them are mediocre. You want to pick out one that will actually teach your employees and help

them to do well when Six Sigma is implemented.
- Double check to see what the return on training investment is going to be. If it is not at least 20 times, then you are going to be wasting money or you may need to find a new project to work on.
- You should start getting the movement going right from the shop level. You do not want to have just a few Green or Black Belts doing the work all the time. You should train the operators and supervisors of the shop floor to use some of the tools and techniques. They can be trained with the White Belt Program to get it done. This helps them to feel some ownership in the process and like they are doing some of the improvements on their own. You can also reward leaders and team members when they get certification to help encourage them to get on board as well.
- Develop a mentoring process: This can make sure that the right guidance is done to ensure that all people are trained. In addition, it can help make sure that course corrections are made on a regular basis and that all projects are done on time.
- Always ensure there is some financial validation of the project. There should be a financial leader who will sign off on how much the project will help the business save money. This can be done during the control phase.
- Make sure that when you are using Six Sigma, you never classify it as the quality manager's job. The quality manager has a distinct role, and they are not going to be in a position where they can manage the process for Six Sigma all on their own. Make sure that the proper team, with the right training, is set up to handle this.
- Create a goal you share in common: Once you have decided that it is time to implement Six Sigma, the next thing that you should do is make sure that all of your

qualified team members are on the same page. This common goal can be shown through an executive directive and it needs to be established for all employees. The point of doing this is to reduce the variability to help reduce waste.

- Standardize the methodology that you want to use: To make sure that your Six Sigma project is going to be successful, you need to have it define a standard approach. If you do not do this, then many individuals on the same team are going to spend their time redefining it again. Standardizing the process in order to reach Six Sigma is going to allow the people on your team to focus on reducing the standard deviation in their own projects, rather than trying to figure out what method they should be using. This standardization may take more time in the beginning, but it ensures there is a common approach, which means execution time of projects can be reduced. In addition, it can create a language that is common, allowing for a true culture of teamwork in the business.
- Map the plan: To make sure that the program you are doing is focused and keeps running on time, you need to make sure that the plan is mapped out completely. You also need to make sure that your teams for each project are identified and that the process is scheduled. The organization needs to be aware that process improvement programs are not going to be implemented within a few days. It can take a few months to a few years to do this. During this time, you must make sure that you invest your resources of money and time wisely.
- Set times to present the data: Throughout the implementation of your new program, there should be frequent reviews and audits. This is done to ensure that

there is some progress being made through implementation. The data needs to be presented and each team needs to be able to describe their milestones, progress, any roadblocks, and the needs and findings that they have.
- Create some methodologies for optimizing any processes that are non-technical: Some invisible processes, such as those that are done by the departments of finance and purchasing, need to be defined, measured, quantified, and optimized. This is true despite the fact that they are nontangible in nature.
- Pick the right project: When you are working on Six Sigma, you are likely to see that there are several projects that you can work on. However, you do not want to waste your time and resources on a project that is not going to make a big difference. You want to work on one that is going to be able to really improve your current process and earn you money. Make sure that everyone is certified properly and able to handle the Six Sigma methodology before you decide on which project you want to work on. This will ensure that you are picking out the right one that can give you the most bang for your buck.
- Go with a project that meets your company goals. There may be many projects from which to choose. However, if you are going to one that does not align with the goals you have for the organization or seems to go against what the organization values, then it should never be picked, regardless of what the Six Sigma process tells you. Make sure that you really implement a project that works with your business, or it is going to end up failing, no matter how hard you work in the process.

When you implement Six Sigma properly, you will be able to reduce any inefficiencies that are there and in return, it is going to produce very high yields for your company. Nevertheless, to make sure that you are as successful as possible, an organization needs to invest wisely, have plans in place, and have some long-term goals so they know where to go next.

Being able to come up with a comprehensive and cohesive strategy when you start implementing the Six Sigma methodology is going to ensure that the company will achieve their business goals.

Conclusion

Thank you for making it through to the end of *Lean Six Sigma: A Beginner's Step-By-Step Guide to Implementing Six Sigma Methodology to an Enterprise and Manufacturing Process.* Let us hope it was informative and able to provide you with all of the tools you need to achieve your goals.

The next step is to start implementing Six Sigma in your business strategy. Many businesses end up producing a lot of waste in terms of money, time, and materials. With the help of Six Sigma, you can learn where these problem areas are for your business and come up with a solution to reduce this waste. This often results in more efficiency and less waste compared to other methods. This guidebook took some time to explain Six Sigma and look over the different steps that you should take to get this methodology started for your needs. When you are ready to reduce waste and help increase your profits and efficiency, look through this guidebook, and learn how to get started with Six Sigma in your business.

Finally, if you found this book useful in any way, a review on Amazon is always appreciated!

Lean Startup

The Complete Step-by-Step Lean Six Sigma Startup Guide

Introduction

Congratulations on getting a copy of *Lean Startup: The Complete Step-by-Step Lean Six Sigma Startup Guide* and thank you for doing so. There are two questions that any company can ask to both reduce unnecessary failure while at the same time ensuring that the company focuses only on ideas that have promising potential. They are:

- Should we build this new service or product?
- How can we improve our odds of success with this new thing?

The Lean method is equally useful for startup companies as it is for Fortune 500 companies. It may have its roots in the technology sector but it is already being used in virtually every industry across the board. While there is lots of confusion around it, the Lean Startup system can help companies of all sizes in a lot of different ways.

While the term "startup" generally has very specific connotations in the business world, in this instance, "startup" simply means any team that is planning to create a new product or service whose future isn't 100 percent certain yet. Generally speaking, it makes far more sense to classify startups as enterprises taking on the challenge amidst uncertainty, than by categories like market sector, size or even age of the company.

With this definition in mind, you will find that there are a few main areas in which a startup faces the greatest amount of uncertainty, otherwise known as risk. Technical or product risk can be summed up by the question "Can it be built?" As an example, doctors who are currently working towards a cure for cancer can be thought of as a startup institution because there is a very large technical risk and this area of study has been going on for quite some time with no hint of success. However, if they do discover a cure, there is absolutely no market risk because its target market would definitely buy it.

Market risk, also known as customer risk, is simply the risk when the product or service reaches the market and no one is actually going to want to buy it. A cautionary example of this type of risk is a company named Webvan that spent millions and millions of dollars creating an automated means of buying groceries online. The only problem is that they tried to get this system up and running in the early 2000s. This is a time when many people were still getting comfortable with the concept of the internet in general but the comfort in buying everyday products online did not follow until nearly a decade.

The business model risk is the risk associated with taking a good idea and building a functioning business plan around it. Even if you already have a good idea, the right business model could very well not be visible until the service is up and running. As an example, when Google started its original business plan of selling advertisements based on previous searches, the plan wasn't clear because no one had done that sort of thing before.

While every company will need to deal with these risks to varying degrees, the biggest risk that most new products or services struggle with is customer risk. It can be difficult to determine the value of something new for customers who haven't experienced it yet. The tricky part here is that in most instances, it will actually appear that the product risk is the most urgent risk. After all, most new ideas don't make it this far without an assumption that someone, somewhere is going to want the product or service at hand. This assumption, then, can lead to a much costlier course of action wherein you do the work to create the product or service before offering it to anyone.

This is where the Lean Startup system comes into play. This technology potentially stops you from being one of the millions of companies out there that has a good idea and a cool product but had crashed and burned because they inherently relied on assumptions about consumer behavior that simply turned out not be true. It is important to think in terms of risk as opposed to company history because in doing so, you will find that many large companies have startup organizations within them. As an example, consider the Gillette razor company who felt that there was little risk in adding the fifth blade to their flagship line of razors because they knew the business model, the market, and the product ins and outs. However, the company that owns Gillette, Proctor and Gamble, operates a startup in the form of its research and development division that focuses specifically on hair removal. With each new idea, this division seems like a startup because they have no known variable which means everything they are working on is extremely risky.

Currently, one of the well-known companies that using the Lean Startup system is General Electric, which is also one of the largest companies in the world. The company has trained more than 10,000 managers around the world to use Lean Startup principles and has used the system to successfully improve the end result on all of their products including refrigerators and diesel engines.

To follow in their footsteps, the following chapters will discuss how to operate a Lean Startup successfully, starting with an overview of the Lean Startup methodology. Next, you will learn how to create a trial startup system that is not only useful but also designed to provide you with as much viable information as possible. You will then learn how to take a successful startup and grow it until it reaches its full potential. From there you will learn about adding Six Sigma and other Lean tools to your startup for maximum efficacy.

There are plenty of books on this subject on the market, thanks again for choosing this one! Every effort was made to ensure it is full of as much useful information as possible, please enjoy!

Chapter 1: Lean Startup Options

While the idea of the Lean Startup has been around since 2011, many companies are still coming to grips with everything the system has to offer. This is despite the fact that most of the ideas presented in this system were hardly new. This is largely due to the fact that the system actually offers more value to established organizations than it does to startups. However, startups can still be able to build a Lean system from the ground up if they choose to.

Lean Startup methodology

Build, measure, and learn: Perhaps more than anything else in recent history, the application of the scientific method to demolish uncertainty, where innovation is concerned, has transformed the way breakthroughs happen. Broken down, this includes the process of defining a hypothesis, creating a prototype to test the hypothesis, testing the prototype (and thus the hypothesis) and then adjusting as needed. While this may seem simple, it has the potential to generate massive results by enabling companies to take risks on smaller ideas without breaking the bank in the process.

The build, measure, and learn approach can be used for virtually everything, not just entirely new ideas. It can be used to test things like customer service ideas, the process of managerial review, or even a new feature for an existing product or service. As long as you can perform a test that clearly validates or disproves the initial hypothesis, then you will be good to go because you must be able to gather enough data to justify approving or vetoing the idea.

The goal, then, is to do everything possible in order to ensure that build, measure, and learn process proceeds from start to finish as quickly as possible. This will make it feasible to run the process multiple times if needed, while also making it clear when such additional runs are needed. As such, it is important to have a very specific idea for each test because as more variables are added, the more difficult it will be to determine results with any real degree of accuracy. When it comes to products and services, this means determining if they are either wanted or needed by the target audience.

Minimal viable product: Generally speaking, most product development involves an extreme amount of work up front. The process involves working through the full specifications of the product, as well as a significant initial investment when it comes to capital in order to build and test multiple iterations of the product. The Lean Startup process thus encourages building only enough of the product in question to make it through a single round of the build, measure, and learn process at a time. This is what is known as the minimal variable product.

The minimal variation of the product is what enables a full cycle of the build, measure, and learn loop to be completed with the least amount of required time and effort on the part of the team. This may not be something as simple as writing a new line of code, it could be an elaborate process that outlines the customer journey, or a complete set of mockups made out of a cheaper substitute. As long as it is enough to test the hypothesis, then it is good to go.

Validated Learning: An important part of the Lean Startup process is ensuring that you are testing your hypothesis with an eye towards the right metrics. Failing to do so can make it easy to focus on vanity metrics instead. Focusing on vanity metrics may make you feel as though you are making progress while not actually telling you all that much about the value of the product. For example, for Facebook, the vanity metrics are the things like the total number of "Likes" that have been received or the number of total accounts created. The real meat and potatoes are in metrics such as the amount of time the average user spends on the service per week. Early on, the metric that validated the company's initial hypothesis was the fact that more than half its user base came back to the service every single day.

Innovation accounting: Innovation accounting is what makes it possible for startups of all sizes to prove, in an objective way, that they are creating a sustainable business. The process includes three steps, starting with determining the baseline. This involves taking the minimum viable product and doing what you can to determine relevant datapoints that can be referred back to the fact. This could involve things like a pure marketing test to determine if there is actually interest from customers. This, in turn, will make it possible for you to determine a baseline with which to compare the initial cycle of the build, measure, and learn process, too. While better numbers are always desired, poor numbers at this stage aren't terribly important, it only means that the team will have more work to do in the build, measure, and learn cycle.

After the baseline has been determined, the next step is going to be to make the first change to determine what can actually be improved upon. While this certainly makes the entire process take longer than it usually does, making too many changes at once makes it difficult to determine which one of the changes led to the biggest improvement. However, if you have a lot of potential changes to test, you can then test them in groups so when something pops, you will only have to retest a specific range in order to see what caused the inspiration to strike.

Once several build, measure, and learn cycles have been completed, the product should be well on its way from moving from the initial starting point to the final, ideal phase. At some point, however, if things don't seem to be proceeding according to plan, then the question becomes whether it is better to pivot to something new or to stick with the current baseline a while longer to see what improves. The choice between the two should be relatively obvious at this point based on the data provided up to this point.

If the decision is ultimately made to pivot at this point, then it can be quite demoralizing for the team because this means going back to square one, albeit with additional data to draw on in the future. Nevertheless, issues such as vanity metrics or a flawed hypothesis can lead teams down a path that is ultimately not viable. This scenario leaves them no choice but to tear it all down and start again with an alternate hypothesis and a clean slate. It is important to try and reframe the idea of a pivot from a failure to a success because it saved the startup from potentially taking a flawed product to market and paying in a big way further down the line.

There are a few additional types of pivots as well. A Pivot that zooms in is one that takes a signal successful feature of a failed prototype and turns it into its entirely own product. A zoom out pivot, on the other hand, is when a failed prototype is useful enough to become a feature on something larger and more complicated.

The customer segment pivot occurs when the prototype proves solid, but the target audience proves to be different than anticipated. A customer need pivot occurs when it becomes clear that a more pressing problem for the customer exists, so a new product needs to be created to handle it.

A platform pivot occurs when a single application becomes so successful that it spawns an entire related ecosystem. A business architecture pivot occurs when a business switches from having low volume and high margins to high volume and low margins. A value capture pivot is one of the most extreme as it involves restructuring the entire business to generate value in a new way. The engine of growth pivot occurs when the profit structure of the startup changes to keep pace with demand.

Small batches: When given the option to fill a large number of envelopes with newsletters before sending them out, the common approach is to do each step in batches, fold the newsletters, place them in the envelopes, etc. However, this is actually less efficient than doing each piece by itself first, thanks to a concept known as single piece flow, a tenant of Lean manufacturing. In this instance, individual performance is not as important as the overall performance of the system. Time is said to be wasted between each step because things need to be reorganized. If the entire process is looking at a single batch, then efficiency is improved.

Yet another benefit to smaller batches is that it is easier to spot an error in the midst of them. For example, if an error was found in the way the envelopes were folded once all the newsletters had been folded, then that entire step would need to be repeated, adding even more time to the process. On the contrary, a small batch approach would determine this error the first time all the steps were completed.
Andon cord: The Andon Cord was used by Toyota to allow any

employee on the production line to halt the entire system if a defect was discovered at any point. While this is a lot of power to give to every team member on the floor, it makes sense as the longer a defect continues through the process, the more difficult and costlier it will eventually take to remove. As such, spotting and calling attention to the problem as quickly as possible is the more efficient choice, even if it means stopping the entire production line until the issue is fixed.

Continuous deployment: Continuous deployment is one of the most difficult Lean Startup processes for many companies to deal with as it means constantly updating live production systems each and every day until they reach an ideal state. The essential lesson is not that everyone should be shipping fifty times per day, but that by reducing batch size you can make it through the entire build, measure, and learn cycle more quickly than your competition can. The ability to learn directly from customers is essential in this scenario as it is one of the primary competitive advantages that startups possess.

Kanban: This is another part of the process that is taken directly from Lean manufacturing. Kanban has four different states. The first of which is the backlog which includes the items that are ready to be worked on but have not yet been actively started on. Next is in progress, which is all of the items that were currently under development. From there, things move to build after development has finished and all the major work has been done so that it is essentially ready for the customer. Finally, the item is validated by a positive review from the customer.

A good rule of thumb is that each of the four stages, also known as buckets, should contain more than three different projects at a time. If a project has been built, for example, it cannot then move into the validation stage until there is room for it. Likewise, work cannot start on items in the backlog until the progress bucket has been cleaned out enough to free up the space. One outcome that many Lean Startups don't anticipate is that this method also makes it easier for teams to measure their productivity based on the validated learning from the customer as opposed to the number of new features being produced.

Five whys: Many technical issues still have a root at a human cause at some point in the process. The five whys technique makes it possible to get close to that root cause from the beginning. It is a deceptively simple plan, but one that is extremely powerful when used by the right hands. The Lean Startup system posits that most problems that are discovered tend to be the result of a lack of personal training, which on the surface can either look like a simple technical issue or even one person's mistake.

For example, with a software company, they may see a negative response from their customers regarding their most recent update. Looking more closely at the issue, it was discovered that this was due to the fact that the update accidentally disabled a popular feature. Looking closer still, this was discovered to be due to a faulty service which failed because a subsystem was used incorrectly due to an engineer that wasn't trained correctly. Looking closer still, you will find that this is due to a fact that a specific manager doesn't believe in giving new engineers the full breadth of training they need because his team is overworked and everybody is needed in one capacity or another.

This type of technique can be especially useful for startups as it gives them the opportunity to determine the true optimum speed needed to make quality improvements. You could invest a huge amount in training, for example, but that doesn't mean this is always going to be the right choice at the given stage of development. However, by looking closely at the root causes of the problems in question, you can more easily determine where there are core areas that require immediate attention as opposed to solely focusing on surface issues.

Another related issue is connected to the fact that many team members are likely prone to overreacting to things at the moment, which is why the 5 Whys are useful when it comes to taking a closer look at what's really happening. There can be a tendency to use the Five Whys to point blame, at first, but the real goal of the Five Whys is to find any chronic problems caused by bad process, not bad people. This is also important to ensure that everyone is in the room together when the analysis takes place because it involves all of the people impacted by the issue, including both customer service and management. If blame has to be taken, it is important that management falls on the sword for not having a team-wide system in place to prevent the issue in the first place.

When it comes to getting started with the Five Whys, the first thing that should be focused on is instilling a feeling of trust and empowerment in the team as a whole. This means being tolerant of all mistakes the first time they happen, while at the same time making it clear that the same mistake should not happen twice. Next, it is important to focus on the system level as most mistakes are made due to a flaw in the system which means it is important to put the focus on this level when it comes to solving problems.

From there, it is important to face the truth, no matter how pleasant or unpleasant it might be. This method may bring up some unpleasantness about the company as a whole but the goal is to fix these issues, after all, and you can't fix what hasn't been brought to light. This is why it is easy to turn it into the Five Blames if you aren't careful which is why the blame should flow up in this instance. Start small and be specific. You want to get the process embedded, so start with small issues with small solutions. Focus on running the process regularly and involving as many people as you can.

Finally, it is important to designate one person on the team as the Five Whys Master. This person will be the one who is primarily in charge of seeing that change actually comes to the team. This, in turn, means they will need a fair amount of authority in order to ensure things get finished. This person will then be the one accountable for any related follow-up, determining if the system is ultimately paying off, or if it is better to cut your losses now and move on. While it can ultimately be a great way to create a more adaptive startup, it can also be harder to get into the groove of than it first appears, so it is important to look at it as a long-term investment rather than something that will be completed in the short-term.

Chapter 2: Create a Useful Lean Startup Experiment

Qualitative or Quantitative: While many people assume that their startup experiment needs to be either quantitative or qualitative, the fact of the matter is that one is not inherently superior to the other. Instead, it is better to think of the two as if one was a hammer and the other was a screwdriver. While a hammer is better at putting nails in wood, that doesn't mean it is inherently superior on all fronts. Any tool can be used for good or evil, which is why it is important to focus more on validating the right metrics than it is to worry about which of these two processes is superior. In fact, using qualitative research and then validating it with quantitative research is likely going to do the most good anyway.

Generative or Evaluative: A generative research technique is one that doesn't start with a hypothesis per se but can still result in a wide variety of different ideas. Things like Customer Discovery Interviews fall under this type of technique. Evaluative, on the other hand, is all about testing a very specific hypothesis in order to determine a very specific result. The popular smoke test falls under this type of testing. It is perhaps this distinction, more than any other, that explains why some people end up with poor results from their experiments.

For example, a smoke test could be run to test the hypothesis that some percentage of the market will be interested in shoes that are compostable. To test this hypothesis, you would then put up a fake coming soon landing page explaining that compostable shoes are totally going to be a thing and see who signs up for the newsletter. After the work was done and the results were in, it turns out that there was about a 1 percent conversion rate when it comes to the shoes. The good news is that the hypothesis was confirmed, the bad news is that it wasn't particularly useful.

What's more, the results are unclear because it still isn't clear if the interest isn't there, if the advertising was poor, or if there is a third variable that you aren't yet aware of. This can be broadly defined as the difference between people not being interested in the value proposition and people not understanding it. The truth of the matter is that there are hundreds of reasons out there why someone might get a false negative result from a given test, just as there are a number of reasons why a false positive might be generated.

To get started, you will need to determine if the hypothesis is flawed or simply vague and, in this case, it is both. Some people are too vague when it comes to a target audience, some are a poor qualifier. As such, first, you would need to focus on a more specific demographic, and second, you would need to do research to determine how big the audience for compostable shoes would ultimately be. Only once the hypothesis is falsifiable and specific can it benefit from an evaluative experiment like the smoke test. If you can't clear up your hypothesis then you will want to start with Generative Research and work back from there.

Market or product: When it comes to the distinction between methods and tools, the biggest is perhaps the distinction between Product and Market. Some methods are useful when it comes to helping startups learn about their customers, their problems, and their best lines of communication. As an example, startups can listen to their potential customers to make it easier for them to understand their specific situations and what their day to day problems are like.

Other methods make it possible to learn about the product or a potential solution that will help to solve a specific problem. One good place to start is with a set of wireframes as a means of determining if the interface is as usable as it seems at face value. Unfortunately, this still won't make it clear if anyone is going to buy anything in the first place.

As these methods don't typically overlap all that well, it is important to choose one and stick with it throughout its cycle. If you combine evaluative research and generative research with Product and Market, you will end up with four different means of determining the best path forward.

Generative Market research asks questions like:

- Who is our customer?
- What are their pains?
- What job needs to be done?
- Is our customer segment too broad?
- How do we find them?

If you can't answer these questions clearly and easily, then your startup is in what is known as the Customer Discovery phase. During this phase, it is important to get to the basis of the problem prior to testing out any potential solutions to ensure that you are actually solving the right problem in the end. If you don't have a clear hypothesis to start, then you will need to generate ideas.

To do so, you may want to talk to customers to see what is bothering them or you could use a data mining approach to determine the problem, assuming you have access to enough data. You may even want to use a survey with open-ended questions if you are really fishing for ideas. Some of these methods will be qualitative and some will be quantitative, but this distinction is ultimately irrelevant in the long run. Data mining is a quantitative approach, but it helps identify problems, most famously the existence of food deserts which would have been difficult to determine in virtually any other way.

Generative Market Research Methods include:

- Surveys
- Focus groups
- Data mining
- Contextual inquiry / ethnography
- Customer Discovery Interviews

Evaluative Market experiment questions include things like:

- How much will they pay?
- How do we convince them to buy?
- How much will it cost to sell?
- Can we use scale marketing?

In order to properly evaluate a specific hypothesis, you may want to start with a landing page to determine if there is likely to be a demand. You may want to put together a basic sales pitch if you are working on a B2B enterprise type product. You could even go so far as to run a conjoint analysis as a means of further understanding the relative positioning of a few value propositions.

Evaluative market experiments that are useful if you have a

clear hypothesis include:

- High bar
- Fake door
- Event
- Pocket test
- Flyers
- Pre-sales
- Sales pitch
- Landing page
- Video
- Smoke tests
- Surveys
- Data mining/market research
- Conjoint Analysis
- Comprehension – link to the tool
- 5-second tests

While this sort of research can provide lots of interesting data, it is important to keep in mind that much of it still has the potential to be wrong as signing up for a landing page is very different than actually putting money down on a product. In any situation where the customer doesn't have to commit anything more than an email address, then they don't signify an actual customer demand.

It is important to keep in mind that the value proposition and the product are not the same things. The value proposition is the benefit that your product will deliver to your target audience. As such, you cannot have a validated value proposition if you don't have a validated customer segment.

Chapter 3: Growing a Startup

When a startup is composed of only a few people, the small team that started the company, it's easy to manage everyone and everything. You've got your first few clients, and they're happy with your work, paying all their bills on time and referring your services to other potential clients. But as your startup grows—with more staff, more clients, and more money to keep track of—it can be a challenge to manage all these aspects efficiently.

But there are ways to make this process easier so that you don't lose too much time or money. It's all about ensuring you use collaboration, effective lead generation, and strict budgeting. Here's how:

Collaboration

It doesn't matter if you're working with 5, 25, or 75 employees; any team, regardless of size, needs to have the right tools and resources to successfully collaborate. Teams, whether they are working alongside one another in the same space or remote, need to have awareness of the initiatives their colleagues are pursuing. Yes, there are many collaborative tools available that allow teams to message one another throughout the day and share files, but what these tools often lack is context.

Cage is a new platform that enables contextual collaboration. Through Cage, teams have the ability to gather feedback in real time, assign tasks, edit images, and distribute media files, all on one platform. By facilitating the entirety of a project, from the initial brainstorm to a final review before a video or platform is published, Cage ensures that everyone involved in the projects has full insight into updates and strategic pivots.

Regardless of the medium, every project takes on a life of its own. More often than not, facets shift over time, and these changes and discussions are often implemented across several platforms, which often lead to confusion and oversight. Cage helps teams avoid this pitfall and, as a result, empowers them to collaborate more effectively and efficiently.

Effective lead generation

All startup founders need to focus on revenue efficiency. Otherwise, according to Forbes, you'll be totally lost. Strategic planning and tracking are the only way that you'll be able to ensure you're being as efficient as possible when it comes to revenue and effective lead generation.

LeadCrunch advises that for every dollar spent on customer acquisition, a company should generate $2.50 in return. But this is not possible if you don't prioritize top-of-the-funnel sales leads. Too many organizations waste time and money by casting large, and irrelevant nets at the top of the funnel, which results in salespeople wasting their time trying to engage customers who, simply, aren't interested. LeadCrunch's CEO, Olin Hyde, believes that the key to successful lead-generation is micro-segmenting, engagement, and nurturing.

Fueled by their AI-driven platform, LeadCrunch allows companies to pinpoint relevant, high-quality prospects, and cultivate deep relationships with them using information that is specific to their unique needs. As a result of their platform's laser-sharp focus, LeadCrunch's top performing customers often see conversion rates spanning 300-1000 percent after leveraging the platform's technology.

Strict budgeting

It's not enough anymore for a startup to use an Excel sheet to calculate and keep track of all their budgeting needs. Even if they've moved onto Google Sheets, which are, after all, free—that's still not enough. If your startup is growing, it makes sense to invest in a B2B budgeting tool that uses actual objective data, which will help you understand how different actions and decisions affect the money coming in.

Hive9, a planning, budgeting, and analytics solution created specifically for B2B marketers, is a smart way for you to keep track of every aspect of budgeting. What's most effective about it? According to Olive & Company, "This comprehensive tool integrates your marketing goals, plan, and budget with your campaigns, so you can measure success and strategically allocate your budget...Hive9 helps you determine where your revenue is coming from, down to a specific touchpoint, and where you can improve your marketing. It also helps determine your cost per marketing lead and sales qualified lead."

According to Hive9's mission, they want to help you create a budgeting plan that you can stick to, one that directly correlates to the complex projects you have going on. They understand how stressful it can be to juggle marketing, which is ever-changing, with budget planning.

Growing your startup is certainly going to be a challenge—but it's a challenge worth taking. After all, change and growth are one of the best ways to ensure your startup thrives and succeeds. By using the strategies of collaboration, effective lead generation, and strict budgeting, you'll be able to improve your startup while also keeping control of all aspects as it grows.

Have you ever grown your startup from its original size? What challenges did you come across, and what strategies did you use to make the transition easier?

Product development

Listen and listen well: Listening is fundamental to a Lean Startup. Listening is what gets you to notice your customers' needs.

It includes listening to devotees and skeptics – your fans will fuel your endeavors and reassure you of all your goodness; skeptics will provide you food for thought on angles perhaps obscured.

Listening will provide you with the "secret to unlock success" and make sense of the noise and valuable feedback that will impact your design iterations, remodeling, and execution.

Educate yourself on laws, regulations, business etiquettes, and many other considerations in your industry.

Your first customers: You must be a die-hard fan of your first customers. After all, they are the ones who were the early adopters, the ones who chose to trust you for the product you introduced to them. And keep selling to them again and again.

Most of the time, startups spend and are more than willing to spend to attract new customers. They forget about their existing customers and do not give importance to re-attracting, reselling, and retaining them.

As a Lean Startup company, your fastest way to feedback for improving your product is the customer who bought it from you at the very start. They tried out your product and knew what was required to make it better for them and many others. Also, your existing happy customers are most likely to spread the word amongst their acquaintances. And referrals are the most authentic and impactful method to increase sales than many other paid programs you are likely to invest in as a startup.

Thus, get feedback from your existing customers and introduce your improved products to them to try out and give you more feedback.

Turn them into loyal customers who will set a precedent of customer value and care for your startup.

Organic outreach: Get to know your customers and let them know you. This is a timeless and evergreen advice for you to grow your Lean Startup faster.

Understand what an SEO audit is and invest in one for your website and social media platforms. An SEO audit will help you create and understand your startup's penetration in the market. Leverage organic outreach and give something of great value to your customers.

A simple combination of great content creation, customer feedback, and input will create a measurable impact on your organic outreach

Content marketing: We are avid fans of content marketing and know that content is here to stay—but only "great" content.

A great content marketing strategy and strategist will make you think of creating resources that are useful to your consumers.

Great content will introduce your business properly, talk about its uniqueness, products, and how people think it is doing.

Your social media and website should have sufficient content, updated timely, and posted regularly to keep things fresh and relevant. Therefore, ensure that you create interesting content making use of the many simple tools available in the market.

A strong online presence with an effective SEO strategy and great content will not only help your startup to attract new customers but will also retain existing ones.

Ask your customers to chip in: A great strategy to grow fast is to involve your customers to build a prototype with you.

The lean methodology begins with a true hypothesis – a fairly accurate guesstimate to solve a problem with a product. Introduce the model to your customers and ask them to give you feedback on the usability of the product.

Feedback: Since Lean Startup companies are closely attached to their customers, you will relatively have more convenient time developing content and resources for your customers and prospective audience. Your business model facilitates a fluid and comfortable relationship with your market. And this close relationship can spark a valuable and insightful market research.

Ask your customers for feedback on what they think you are doing regarding selling your products, providing support, quality management, and areas for improvement. Listen carefully and make sure you create a list of their feedback and ideas.

Customer onboarding and satisfaction

As a startup, know your customers personally. Create opportunities to meet them. Many small business brands send handwritten notes with their products to customers to make them feel welcome and part of the family. Send birthday cards, season greetings, etc. to engage your customers.

Share stories of growth, challenges, and outcomes: Your content marketing strategy must include the changes you make, big or small, for example, a new and improved design or a slight color variation, or updating a new phone number to manage queries to the addition of new members in your sales team, etc.

Tell your customers when you make changes to your products based on their feedback, as it will create a circle of trust and reliability which will ultimately affect your bottom line.

Sharing information this way will establish a culture code in your startup and communicate your value proposition via meaningful actions to both your customers and prospective customers.

Accept mistakes: There will be mistakes made on the way. Own them and seek to clarify and apologize appropriately.

Paid outreach

At times, paid marketing and outreach can work brilliantly for a startup. However, you will need to think through before executing this strategy.

The sure sign of a well-paid outreach is profit. You should be making more money with paid outreach per customer than you would in acquiring one. Even if the ratio is 2:1, paid outreach is working.

By the end of the day, your goal is to make customers happy. When your customers are happy with your product, they will refer you to their friends and family, and you will continue to succeed in your endeavors.

And this leads us to our suggestion of looking into influencer marketing. Influencer marketing will involve your happy customers to become your brand ambassadors and talk about your product in their social circles to co-build your brand.

It is a win-win for you, your existing customers, and your potential customers who will have been influenced by their friends or influencers. This way, your startup's value proposition travels smoothly into the market.

Build a great team

Keep good people around you. Keep people who know what they are doing. Keep people who thrive on feedback and know how to use that feedback for growth and improvement; this is key to the interactive model of improving your product and keeping it relevant to your customers' needs.

Learn from examples: These are some examples of well-known companies who did not want to waste their time, their customers' time, or that of their investors. They got down to developing the product that would fulfill the needs of their customers.

When Dropbox came across the lean methodology, the

company began thinking about their product which at the time of its inception was competing with many other similar products.

They learned how to test new products with their customers and incorporated their feedback. From 100,000 registered users, Dropbox went over 4,000,000 in only 15 months with the Lean Startup principles.

The giant General Eclectic also went into the Lean methodology to develop faster solutions with FastWorks - a complete mindset changing program to how things usually worked at GE. A very popular example was the development of a gas turbine two years faster with 40% less cost. The lean methodology is not only for startups after all!

After approaching local shoe stores and grabbing photos of their shoes with consent, the owner of Zappos tested his hypothesis of whether an online shoe-selling site would work. And it did! Zappos is the largest online shoe store with over 1000 brands featured.

The ideas above are fairly consistent with the ideas you hear every day. You learned these at business schools, through books and blogs, and friends and colleagues. Simple as they may sound, it is always the simpler ideas that are key to great outcomes.

Chapter 4: Six Sigma Basics

Six Sigma is the name of a lean system for measuring quality with the ultimate goal of getting as close to perfection as feasibly possible. Ideally, a company that is running at Six Sigma maximum efficiency would generate, at most only 3.4 defects per million attempts at a particular process. Zshift is the name given to these deviations which show the difference between a poorly completed process and a perfectly completed one. The baseline Zshift is 4.5 while the ideal value is 6. Processes that have not yet been analyzed via the Six Sigma process typically score somewhere between 1 and 2.

Levels of Zshift: If the Six Sigma analysis of a process is at 1, then this means you can expect customers to get exactly what they want somewhere around 30 percent of the time. If this is increased to a Zshift 2, then you can expect the process to give the customers exactly what they are looking for about 70 percent of the time. If the process reaches a 3, then it will be accurate about 93 percent of the time, a 4 will be 99 percent accurate, and a 5 and 6 reach even smaller divisions towards the goal of 100 percent accuracy and customer satisfaction.

Six Sigma cert: Within Six Sigma, there are a variety of different certification levels that can be achieved, each with its own tasks and responsibilities as it relates to the whole. Six Sigma is all about decreasing the risk of production errors by reducing waste and improving efficiency. The first two levels of certification, White and Yellow Belts, are crucial to this part of the process as, while working under higher level Sigmas, they do things like ensure the data that is coming in is on the right track and carry out specific functions that are designed to add overall value to the process. These certificates are also a great way to be exposed to the overall Six Sigma methodology.

The next level of certification is the Green Belt which allows holders to work more directly on Six Sigma projects being helped by those above them while also allowing them to oversee projects being handled by Yellow and White Belts. Black Belts then lead high-level projects while also supporting and monitoring those at other tiers. Finally, Master Black Belts are those who are often brought in specifically to start Six Sigma at a company and are knowledgeable to mentor everyone at every level.

Six Sigma is the tool for you if…

While almost any company and any team can benefit from the Six Sigma in some shape or form, this doesn't mean it is always going to be the right choice, especially for a startup company. In fact, deciding if it is the right choice or not actually depend on a wide variety of different specifics, including how committed the team is to implementing the system in the first place and what the company's developing culture is like.

Leadership involvement is key: When discussing the idea of transitioning to Six Sigma, it is important to not look at it as another type of flash in the pan management style that is going to go out of fashion as quickly as it appeared. Instead, it is far more likely to find purchase if it is pitched as an enhancement of an existing system. Likewise, you can always find opposition to something new from someone in the company which is why it is important to start with buy-in from the top and work your way down the list from there. It is extremely important to have the full management team on board from the start as if it doesn't look like there is a consensus regarding the new system, then it is going to be dead in the water before it even gets off the ground.

This doesn't mean that the entire team needs to be committed to the idea of Six Sigma from the start, but it does mean that it is vital that the change that is on the horizon that needs to be institutional which means the leadership needs to put forth a united front. Don't forget, the human brain is a creature of habit which means that it will recoil from new systems that seem too complicated if the system is, at all, perceived as optional. As such, if there is an opposition to the new system, stress the idea that it is important it be expressed in private.

Consider the infrastructure: Six Sigma is all about leaders mentoring those on their teams in order to make Six Sigma work as effectively as possible. As such, if you hope to transition to a Six Sigma system successfully, then this needs to be a full-time job for at least one person, at least until the new, positive, habits have formed for good. While this might not seem cost-effective in the short-term the Six Sigma savings that will appear once the system is up and running properly are sure to more than make up for it.

Consider what will boost compliance: After support of management has been assured and the infrastructure is in place to make the project really pop, the next step will be to ensure that the rest of the team members have a motivational reason to fall in line. While active rewards aren't going to be required once the Six Sigma process has been properly internalized, they are a good way to help the team get used to looking at problems as though Six Sigma is the solution. While the right way to track team progress is going to be different for every team, it is vital that each member of the team feels an immediate and compelling reason to commit to the new program, at least at first.

After all, companies are like any other body that is currently in motion, the bigger the company, the more inertia is needed when making large changes. This is where your startup has the opportunity to outmaneuver the big boys and start gaining some ground as quickly as possible.

Who else is using Six Sigma in your field: While Six Sigma has proven its worth in a wide variety of different fields, this doesn't mean that each of these fields is going to be ready to adopt the process with open arms immediately. While looking to the future is one thing, if your startup is also going to the first company to adopt Six Sigma as a common practice in its industry, then you need to be prepared from extreme resistance from every side and especially any members of the old guard that you may have brought on to ensure the real work gets done. Luckily, the science behind Six Sigma is solid which is why it should be fairly easy to come up with specific examples of how it can help your company to silence any opposition.

Consider training objectives: Depending on the size of your team, training everyone as a whole might make sense. Eventually, however, different training levels are going to need to be enforced, as not everyone is going to need to be a Black Belt. As such, it is important to look more closely at the various levels and the qualifications for each before determining how training can best be broken up for maximum effectiveness. It is important to also factor in how the training will affect any other duties the trainees might have as well as what areas are going to be focused on most stringently.

Taking the time to work out the specifics of your training scenario before you get started is sure to make all the difference in the world when it comes to implementing Six Sigma successfully. Remember, there are no one-size-fits-all options in this scenario which means planning out the specifics of your team's training could literally be the success and failure of the entire project. What's more, it can also make other issues more apparent when otherwise, they would not have been noticed until training was already underway.

Consider flagship projects: After the Six Sigma training is out of the way, it is important to have a few important projects waiting in the wings in order to show the team that the system is worth it. Not only will these projects help to get the entire team excited about Six Sigma, but they will also be useful down the line if questions as to whether or not Six Sigma is worth keeping start popping up as well. As a general rule, you will want to start with at least a Green or Black belt project and then do everything in your power in order to ensure they end up being successful.

While doing so, it is also important to not spread the Black Belts and Green Belts that you have on your team too thin that it makes them struggle to match their deliverables. Instead, it is better to have too many people on a few projects to ensure they are completed to perfection. Remember, if things go according to plan then you will have plenty of time to complete other projects once these go over like gangbusters.

Furthermore, it is important that these early chapters are more than just fluff projects, they need to be things that are legitimately beneficial to the company as a whole. If your early projects are heavily publicized but do little to produce viable results, then you run the risk of Six Sigma being seen as little more than a fad with lots of flash and little substance. You can ensure that this doesn't happen by instead choosing projects that have a clear value, regardless of whether the person making the decisions is trained in Six Sigma or not.

Remember, public opinion is one of the most important resources to covet at this stage and will continue to be so until Six Sigma has become a habit for the entire team.

Chapter 5: Implementing Six Sigma

Give the team a reason to want to try Six Sigma: In order to ensure that Six Sigma is implemented successfully, it is vital that you take the time to motivate your team in the most effective way possible, so they understand why it is so important to adopt the Six Sigma methodology. Depending on the state your startup company is in, the burning platform scenario might be the best choice.

The burning platform is a type of motivational technique whereby you explain that the situation your company finds itself in is just as perilous as standing on a burning platform and the only way to turn things around for good is by implementing Six Sigma. It is important to have stats that back this idea up, though fudging the numbers for productivities sake might not hurt either. Adapting to Six Sigma can be difficult for team members who are set in their ways and external motivation may be just what the doctor ordered.

Give team members the tools they need: After the primary round of Six Sigma training has finished, it will be important to ensure that you have a strong mentorship program in place, along with details on the finer points of the process for those who need them. One of the worst things that can happen at this point is for a team member to express an interest in the program only to become disinterested when additional materials are not readily available. A team member who cannot easily find answers to their questions is a team member who will not follow Six Sigma processes when it really counts.

Prioritize properly: Regardless of the situation, there are always going to be a variety of potential outcomes. While talking and planning for Six Sigma is one thing, taking steps to actively prioritize it is another entirely. When team members see those in leadership roles prioritize Six Sigma outcomes, it makes them more likely to prioritize Six Sigma activities in their own jobs as well. Additionally, it is important to make it clear that quality is critical, as is listening to the customer when it comes to ensuring Six Sigma leads to measurable results so that team members of all level of certification can keep an eye on the overall progress the company is making.

Make it a group thing: When it comes to tutoring your team about Six Sigma, it is important to ensure that they make personal connections with how it will affect their jobs for the better so that they feel more personally invested in the program's overall success. This may come about by ensuring the entire team is able to provide buy-in or making different team members responsible for enforcing different aspects of the Six Sigma process as taking the time to ensure that everyone feels connected to seeing Six Sigma succeed will ensure personal retention rates remain as high as possible.

Track the results: Determining a realistic metric that can determine appropriate levels of success before and after Six Sigma is an important step in the process as it can provide you with the motivating data that is required to ensure that Six Sigma adoption is at an all-time high. On the other hand, if it turns out that the system actually ends up being ineffective, then you will be the first to know as well. Regardless, having a viable metric to properly determine aptitude is sure to come in handy more than once. What's more, assuming the results are positive then it is sure to be a great motivating factor for the entire team and provide yet another reason why sticking with Six Sigma is so important.

Reward team players: While offering viable reasons for the team to adopt Six Sigma is one thing, it is still important to provide positive reinforcement during the early days so that everyone is constantly motivated to follow the Six Sigma process until it becomes a habit. The goal here should be to choose a reward that is valuable while at the same time not being so extravagant that eventually removing it won't completely remove the team's desire to keep up the good work.

Six Sigma criticism

Six Sigma is just a fad: While it has only been back in the spotlight for a few years, the fact of the matter is that the origins of Six Sigma can be traced all the way back to the early 1900s when it was used by entrepreneurs like Henry Ford, Edward Deming, and Walter Shewhart. Additionally, it further separates itself from true fad management styles by being more focused on the use of data as a means of ensuring the best decision is made at the moment as possible, specifically those with a focus on the customer as a means of ensuring a viable return on every investment.

Switching to Six Sigma is resource intensive: While it's true that training the team in the Six Sigma process is time-consuming, the end goal is for it to save far more time than the training will cost in the long run by ensuring team members do their jobs as effectively as possible moving forward.

It is important to remember the story of the pair of lumberjacks who worked day after day in the forest. One man worked himself to the point of exhaustion every day while the other man spent the time preparing properly, and at the end of the day, both men had always chopped the same amount of wood. If your team has the opportunity to work smarter instead of harder, why wouldn't you provide them with the tools they need to make that the new norm.

Furthermore, the cost of training the team the Six Sigma process can be further mitigated over time by spreading out the training courses as required. While this means the team won't start seeing the results as quickly as might otherwise be the case, even getting the entire team up to Yellow Belt will produce noticeable results. Furthermore, any funds put towards this type of training can really be seen as an investment in the business as a whole and should be treated accordingly.

Our team is too small for Six Sigma to be effective: While the effectiveness of Six Sigma is proportionate to the inefficiency of the processes previously in place, that doesn't mean it doesn't have something to offer companies that are just getting up and running as well. After all, Six Sigma offers a different way of looking at the types of business interactions that happen day to day in hopes of increasing productivity and, as a result, profits as well, regardless of the size of the team that is utilizing it. What's more, smaller teams will actually be able to take on the Six Sigma mantle more easily than larger companies as the number of resources required to train 10 team members are always going to be far lower than what it would cost to train 50 instead.

Furthermore, smaller businesses can be hindered more by production bottlenecks which means a Six Sigma system would potentially be able to lead to greater periods of growth as issues that may not otherwise have been addressed for years are taken care of before they become institutional problems. Regardless of the size of the company in question, taking the time to truly streamline relevant processes or improve customer relations is always going to be the right choice.

Six Sigma doesn't apply here: While it's true that Six Sigma isn't going to apply to every single industry across the board, it has moved beyond its manufacturing sector roots. Furthermore, studies show that industries that provide services are prone to more waste than the manufacturing sector in the first place. This is due to the fact that so much of what is produced is intangible that it makes standardizing any process extremely difficult. This is where Six Sigma comes in as it has plenty of processes in place to track the services that are being provided which can ultimately be used to improve efficiency.

Six Sigma is difficult to use practically: While it may have a reputation for being all about the numbers, a vast majority of the tools and principles that are used in implementing Six Sigma require less math and more common sense. As an example, consider the mitigation of waste which is one of the most important aspects of Six Sigma, and something that only requires an understanding of the business in question and how it can be done in a more effective fashion. This is indicative of most of Six Sigma which is largely about fostering the mentality that makes it possible for team members to find the root cause of an issue, regardless of how long it might take. While formulas and mathematical equations may be used, they are simply a justification for this fact.

Lean is plenty for now: When expressing the benefits of Six Sigma, it is important to make it clear that it is a variation of the Lean system, not a replacement for it. In fact, the system is often referred to as Lean Six Sigma. When used in conjunction with one another, Lean will then the throughput and speed of the process and simplifying to ensure that the team is able to do the best with what they have available. Six Sigma then takes these improved processes and makes them of the highest quality possible by reducing defects and, as a result, lowers the deviation. Combining the two can only lead to better results overall.

Chapter 6: Additional Strategies

Kaizen

The word Kaizen translates to "continuous improvement" which is obviously an important goal when it comes to creating the most effective Lean system possible. The goal of the Kaizen strategy is to ensure that all of the talent within the team is always focusing on improving whenever and wherever possible. This strategy is relatively unique for a Lean strategy in that it is more than just a direct plan of action, it is also a general philosophy for the company as a whole. The goal of Kaizen is to create a culture that is supportive of improvement in all of its forms while also creating groups that are focused directly on improving key processes and reaching well-defined goals.

Kaizen is a great strategy to implement while you are standardizing your work process as the two complement one another well. Standard practices lead to current best practices which Kaizen can then improve upon. The Kaizen strategy can be useful when it comes to improving any strategy that your team uses with any real degree of regularity as long as you are fully aware of the end goal for the updated process. From there, you will need to review the current state of things before adding any improvements. From there it is just a matter of following up properly in order to ensure any proffered improvements work as expected.

Training the team in Kaizen actually serves double duty as it teaches them to apply the philosophy and the plan of action at the same time. This type of thinking is often habitually formed by those who are constantly looking for ways to improve their most commonly used processes while also allowing other team members to approach common tasks in new and innovative ways as required. This mindset should naturally be nurtured whenever possible as it is the only way to ensure more fruitful results in the long run.

While constantly improving existing practices is a great place to start, it is important that the Kaizen your team is practicing does not only occur after the fact. When new processes are created, it is best for everyone that they are held to the same examination process as any other. Hindsight is useful, foresight gets results.

Kaizen steps
- The first thing you will need to do is to standardize your process, not just the process that you are looking to put through the Kaizen process but all the processes to ensure that any eventual improvements are as beneficial as possible.

- Next, you will need to compare the processes at play in order to determine where steps that are being used in some processes can be used successfully elsewhere as well. When taking this step, it is vital that you look at true KPIs as opposed to anecdotal information for this step as, otherwise, it can be easy to get off on wrong track without even realizing it.

- After you have determined where real change should occur, the next step is to consider what you currently have available to make completing the process as easy as possible. During this period, you will want to consider the start of the project as well as its conclusion and then brainstorm all the possible ways to reach point B from point A. While no idea should be off the table at first, it is important that you ensure you only move forward with ideas that are truly useful as well as innovative as innovation for innovation's sake is only going to create waste.

- The final step is going to be to repeat as needed so that new innovations become standard operating procedures so that you can then begin the entire process anew. When it comes to Kaizen, the only bad idea is to rest on your laurels.

Creating a Kaizen mindset: Getting the entire team together for a Kaizen event where everyone brainstorms ways to streamline a specific process is relatively straightforward. However, training your team to always work from a Kaizen mindset can be a far more difficult task. Difficult does not mean impossible, however, and the best way to start to train them to this improved way of thinking is to focus on creating a corporate culture where elimination of waste is everyone's top priority. If you can keep this idea in the team's mindset, during every meeting, every performance review, every informal conversation, day in and day out, then eventually team members will start noticing waste without even having to consciously think about it. Once this occurs they will be well on their way to finding ways to work around it instead.

With this done, you will also want to start to set aside a specific time each day to allow team members to look at the processes they use every single day and do nothing else but really think hard about them. It is important to always remember that the human mind loves repetition almost as much as it loves patterns which is why it is so easy to follow the steps for a process you have done a hundred times without even thinking about it. While this can make the process go faster if the steps involved are optimized, it can also make it easy to complete the process with blinders on and not notice points of inefficiency while you are in the midst of them. As such, providing the team with the time they need to think about their processes separate from actually doing them will let them look at the entire project from a different angle.

If you take this exercise a step further, you will then provide the team with an opportunity to talk to the rest of the team about their processes as well. This cross-pollination of ideas will give each process an entirely fresh set of eyes which will provide insight into even more blind spots. This is especially useful for particularly complex processes, just make sure that everyone takes detailed notes, so nothing gets lost in the shuffle. Additionally, it is important to emphasize that there are no wrong answers during this stage, a free and open dialogue can provide solutions to problems that you previously weren't even aware you were facing.

Poka-Yoke

The Lean system strategy known as Poka-Yoke is most accurately translated as actively guarding against mistakes. Essentially, Poka-Yoke can be thought of as a variety of failsafe procedures that are naturally built into any processes specifically for the purpose of catching common errors. Poka-yoke is best used on tasks that are especially repetitive, require precise repetition across numerous steps, or require an extreme period of focus to use correctly. This tool is an especially beneficial type of Muda that works to ensure the overall value while not necessarily creating any of its own.

When Poka-Yoke is at its most effective, it relies on a thorough understanding of the steps in every process as well as additional ways of mitigating potential pain points as effectively and cheaply as possible, while also taking care not to create any new bottlenecks as a result.

Control Poka-Yoke: Control Poka-Yoke does not allow the next step of the process to move forward until a found error has been corrected. As an example, the way a USB dongle is designed so that you cannot plug it in unless it is facing the right way is a Control Poka-Yoke as it ensures you cannot plug in the device in such a way that it will not work once you do so.

Warning Poka-Yoke: Warning Poka-Yoke, as the name implies, provides the team member completing the process that they made an error on a proceeding step.

Contact method Poke-Yoke: The contact method Poka-Yoke works under an assumption that a third party, either a device or a person, is monitoring the steps that are being taken to ensure no errors materialize. A spell-check program is a good example of this type of Poka-Yoke. Contact method Poka-Yoke is especially useful if the same task needs to be repeated as quickly as possible in order for the process to run smoothly.

In order to determine where Poka-Yoke can be of the most use to your processes, the first thing you will need to do is to determine which steps in the process already cause the most harm, or have the most potential to cause harm, in the shortest period of time overall. You may want to start by determining the processes' critical features and then looking at potential causes of failure before determining a signal method that will work effectively in the situation.

Fixed Value Poka-Yoke: This type of Poka-Yoke is useful in situations where the overall process is quite short but requires the same step in the process to be run numerous times in a row. Poka-Yoke is useful in scenarios like this as they allow the person who is completing the process to know how many times they have repeated a specific step. As an example, think of an administrative assistant making numerous copies of a document who first counts out the amount of paper they need to ensure they don't have to count each as it is made.

Motion step Poka-Yoke: This type of Poka-Yoke is useful in situations where a team member needs to perform numerous different tasks, in a specific order, many times. This Poka-Yoke determines when specific steps have been completed to ensure the team member completing the process remains on track. As an example, consider any website where you are asked to enter your payment information. When the website tells you that you haven't entered the correct payment details and can also track that you haven't yet checked the box to prove you aren't a robot, then it is an example of motion step Poka-Yoke.

Self-Check Poka-Yoke: This type of Poka-Yoke requires the fewest additional resources to complete and instead just requires a little extra time to give the team member performing the process the opportunity to check their work before they move on to each new step. This is a good choice for scenarios where mistakes are extremely obvious, and its biggest drawback is that it requires extra time during which the team member must remain focused as well.

Task Poka-Yoke: This type of Poka-Yoke is useful when it comes to processes that require a team member to directly come into contact with a customer as it helps cut down on mistakes that are made in live situations. A great example of this is the change machines at grocery stores that prevent cashiers from making mistakes when counting out change.

Treatment Poka-Yoke: This type of Poka-Yoke works to ensure that the customer always has a positive and efficient interaction whenever they encounter a team member while working through the course of a specific process. The goal here is to standardize what team members say as much as possible in an effort to prevent any potential mistakes before they happen. This is especially useful for new businesses as it gives new team members something to fall back on when they encounter something new, which is basically everything at this point. A great example of this type of Poka-Yoke is the scripts call centers use.

Tangible Poka-Yoke: This type of Poka-Yoke aims to standardize the physical element of the customer's experience. In situations where individual customers have widely varying needs, this type of Poka-Yoke is often the best way to standardize service. A good example of this type of Poka-Yoke is uniforms.

Preparation Poka-Yoke: This type of Poka-Yoke: aims to work with the customer directly to influence expectations and goals prior to the experience. Depending on the requirements surrounding the products or service in your business, this can be a great way to make sure that customers are prepared properly before they speak to a team member to ensure the whole process goes as smoothly as possible. A good example of this type of Poka-Yoke are menus that are visible to patrons of fast food restaurants before they order so the actual order process proceeds as smoothly as possible.

Conclusion

Thank you for making it through to the end of *Lean Startup: The Complete Step-by-Step Lean Six Sigma Startup Guide*, let's hope it was informative and able to provide you with all of the tools you need to achieve your goals. Just because you've finished this book doesn't mean there is nothing left to learn on the topic; expanding your horizons is the only way to find the mastery you seek. Additionally, it is important to keep in mind that, while there is some overlap between any two startups, much of what is going to take place is going to be largely unique to the startup in question.

After all, isn't the point of a startup to do something new? As such, it is important to understand that while following the Lean Startup strategy can certainly lead to success, sometimes you may have to make your own way because what you are trying to do is so out there that the existing methods of testing don't apply. This doesn't mean that you should abandon all that the Lean way of doing things has done for you thus far, it simply means that you will need to take what you have learned and use that to create logical ways to test whatever it is you are prototyping. Likewise, it is important to not get impatient and try to rush the process. After all, creating a viable product or service that a targeted portion of the audience is interested in is a marathon and not a sprint which means the slow and steady wins the race.

Finally, if you found this book useful in any way, a review on Amazon is always appreciated!

Lean Enterprise

The Complete Step-by-Step Startup Guide to Building a Lean Business Using Six Sigma, Kanban & 5s Methodologies

Introduction

Congratulations on getting a copy of *Lean Enterprise: The Complete Step-by-Step Startup Guide to Building a Lean Business Using Six Sigma, Kanban & 5s Methodologies*. These days, it is more difficult than ever to build a business that can remain competitive in a world where customers can find your competition with just the click of a mouse. While there is only so much you can do when it comes to adjusting your profit margins, you can still find success by adjusting the method that will complete the processes in making your business successful.

Making a business Lean can give it the competitive advantage that the perpetual buyers' market takes away. However, it can be more difficult than it first appears which is why the following chapters will discuss everything you need to know in order to turn your business into a Lean mean fighting machine. First, you will learn all about the Lean system, its many benefits, and how you can get started creating your very own Lean system. Next, you will learn how to move the process forward in the right way by ensuring that you have the right goals in mind and that you go about implementing them in the best way possible.

From there, you will learn how to create a value stream map which is vital when it comes to ensuring that your business's various processes are truly on point before learning how to choose the Lean system that best supports the flow of production. You will then learn about the importance of standardization before learning about the several important Lean tools which you can use to really whip your business into shape.

There are plenty of books on this subject on the market, thanks again for choosing this one! Every effort was made to ensure it is full of as much useful information as possible, please enjoy!

Chapter 1: Why Lean Matters to Your Enterprise

Lean principles were first discussed by an MIT student by the name of John Krafcik in his master's thesis. Before starting at MIT, Krafcik had already spent time as an engineer with both Toyota and GM, and he used what he learned from the Japanese manufacturing sector to posit a number of standards that he believed could help businesses of all shapes and sizes operate more efficiently.

The basic idea is that, regardless of what type of business a business is in, it is still just a group of interconnected processes. These interconnected processes can be categorized such as primary processes and secondary processes. The primary processes are those that directly create value for the business. Meanwhile, secondary processes are necessary to ensure the primary processes run smoothly. Regardless of the type of process you look at, you will find that they are all made up of a number of steps that can be carried out in a way that ensures they work as effectively as possible and that they need to be viewed as a whole in order for an effective analysis to be completed.

As a whole, you can think of the Lean process as a group of useful tools that can be called upon to identify waste in the current paradigm either for the business as a whole or for its upcoming projects. Specific focus is also given to reducing costs and improving production whenever possible. This can be accomplished by identifying individual steps and then considering the ways they can be completed more effectively. Some tools that are commonly used in the process include:

- 5S value stream mapping
- error-proofing
- elimination of time batching
- restructuring of working cells
- control charts
- rank order clustering
- multi-process handling
- total productive maintenance
- mixed model processing
- single point scheduling
- single-minute exchange of die or SMED
- pull systems

Beyond these tools, Lean is also comprised of a number of principles that are loosely-connected around the twin ideas of the elimination of waste and the reduction of costs as much as possible. These include:

- flexibility
- automation
- visual control
- production flow
- continuous improvement
- load leveling
- waste minimization
- reliable quality and pull processing
- building relationships with suppliers

When used correctly, these principles will ultimately result in a dramatic increase in profitability. When given the opportunity, the Lean process strives to ensure the required items get to the required space in the required period of time. More importantly, however, it also works to ensure the ideal amount of items move as needed in order to achieve a stable workflow that can be altered as needed without creating excess waste.

This is most typically achieved via the tools listed above but still requires extreme buy-in at all levels of your organization if you ever hope for it to be effective in both the short and the long-term. Ultimately, the Lean system is only going to be as strong as the tools your company is using to implement it, and these tools will only ever be effective in situations where its values are expressed and understood.

Important principles

While it was originally developed with a focus on production and manufacturing, Lean proved to be so effective that it has since been adapted for use in virtually every type of business. Before adopting the Lean process, businesses have only two primary tenants. The first one focuses on the importance of incremental improvement while the other one is the respect for people both external and internal.

Incremental improvement: The idea behind the importance of continuous improvement is based on three principles. The first is known as the Genchi Genbutsu and is discussed in detail below. The second is known as Kaizen, and it has its own chapter later in this book. Finally, in order for continuous improvement to be truly effective, it is important to understand that you must lead your business with a clear knowledge of the challenges you are most likely to face as it is the only way to determine how to deal with them effectively.

When doing so, it is important to approach each

challenge with the appropriate mindset which is one that supports the idea that every challenge leads to growth, which, in turn, leads to positive progress. Finally, you will also want to ensure that you take the time to challenge your preconceptions regularly as you will never know when your business might end up operating on an assumption that is no longer true. This is ultimately the best way to find unexpected waste which will ensure that you really start to improve internally not just in the short-term but in the long-term as well.

Respect for people: This tenant is both internal and external as it applies equally to your own people as well as to your customers. Respecting the customers means going the extra mile when it comes to considering their problems and listening to what they have to say. When it comes to respecting your team, a strong internal culture that is dedicated to the idea of teamwork is a must. This should further express itself in an implied commitment to improving the team as a means of improving the company as a whole.

Getting an edge

Prior to the digital age, businesses could determine their sales margin by starting with all the relevant costs, adding on a reasonable profit margin and calling it a day. Unfortunately, the prevalence of screens in today's society means that everyone is a bargain shopper, simply because it takes so little effort. This, in turn, means that you are not only competing against other businesses in your city, county or state, you are now competing with businesses all around the world as well. As such, there are only a few options when it comes to squeaking by with any profit margin whatsoever. Companies can either add additional real or perceived value, or they can reduce the amount of waste they are paying for as much as possible.

Most businesses find that it is better to determine their margins by looking at what customers are likely willing to pay for specific goods or service and then working backward from there. Ideally, you will be able to reduce that price by five percent to ensure you are truly competitive in a cost-conscious world. While it might not seem like much, this extra five percent is extremely important as customers are constantly on the lookout for the next sale, regardless of how much is actually being saved. The mental benefits that come along with being five percent better than those around them will be more than enough for them to commit to your product or service over all the rest.

Value add: Regardless of what your business does, you will find that there are Lean principles that can be implemented in order to improve the overall amount of value you are providing for your customers while also showing them you appreciate their business and respect them as individuals. What's more, you will be able to address the potential for waste in your organization while also maintaining flow and work to achieve perfection.

Often, you can manage this by doing something as simple as listening to your customers' specific wants and needs which will make it easier for you to determine what they really value the most when it comes to the niche your company habits. Value is most often generated by adding on something tangible that either improves or modifies the most common aspects of the good or service being provided. The goal is that this improvement is something the customer is willing to pay for, so when they receive it for free, they see this as a viable reason for your service to cost more out of the gate. It is also very important that the added value is very easy for the customer to claim because otherwise, they will feel that you have misled them.

Cost reduction: As the Lean system is already quite big on cutting down on waste in all of its forms, it should come as no surprise that it has some ideas when it comes to cost-reducing measures. For starters, it is important to understand that when it comes to Lean, all the different types of waste can be broken down into three types.

Muri is the name for the waste that forms when there is too much variation within common processes. Muda is the name given to seven different types of waste including:

- Transportation waste is formed when parts, materials or information for a specific task are not available because the process for allocating resources for active products isn't where it should be.
- Waiting waste is created if a portion of the production chain has ideal time when they are not actively working on a task.
- Overproduction waste is common if the demand exceeds supply, and there is no plan in place to use this situation to the business's advantage. The Lean systems are designed to ensure that this number reaches zero so supply and demand are always in balance with one another.
- Defective waste is known to appear when some segment of the standard operating process generates some issue that needs to be sorted at some point down the line.
- Inventory waste is known to appear if the production chain ends up remaining idle between runs because it doesn't have the physical materials needed to be

constantly running.
- Movement waste is generated when required parts, materials or information needs to be moved around successfully to complete a specific step in the process.
- Additional processing waste is generated if work is completed that does not generate value or adds value for the company.
- A commonly added eighth muda is the waste created by the underutilization of your team. This can occur whenever any member of the team is placed in a position that doesn't utilize them to their full potential. It can also refer to the waste that occurs when team members have to perform tasks for which they have not been properly trained.

Muda also comes in three categories, the first of which is muda that doesn't directly add value but also cannot be easily removed if the system is going to continue working properly. When faced with this muda, the goal should be to work to minimize it slowly as a precursor to removing it completely. The second type of muda is that which has no real value, whatsoever, and you should work to remove it directly once you've become aware of it. Finally, the third type of muda doesn't directly add value but is required for regulatory purposes of one type or another. While it may be annoying, this type of muda is unavoidable in most instances and the best that you can do is ensure you are always updated to any relevant policies.

Chapter 2: Creating a Lean System

Lean leadership

With so much emphasis placed on improving efficiency, the Lean process naturally puts a lot of emphasis on team leaders who should be working hard to directly inspire their teams to adopt the Lean mindset. In the end, many Lean systems live and die by the leadership involved, which means it is important that those who are put in charge of leading the Lean transition are able to not only explain what's going on but are truly committed to the work that is being done as well. Some of the things that Lean leaders should strive to emphasize include:

Customer retention: When it comes to customer retention, Lean leaders need to take the time to consider not only what their customers want at the moment, but what they are likely to want in the future as well. Additionally, it is important to understand what a customer will accept, what they will enjoy, and what they will stop at nothing to obtain. The Lean leader should also work to truly understand the many ways the specific wants and needs of their target audience throughout the customer base.

Team improvement: In order to help their team members be their best, Lean leaders should always be available to help the team throughout the problem-solving process. At the same time, they are going to need to show restraint and refrain from going so far as to take control and just do things on their own. Their role in the process should be to focus on locating the required resources that allow the team to solve their own problems. Open-ended questions are a big part of this process as they will make it possible for the team to seek out a much wider variety of solutions.

Incremental improvement: One of the major duties of the Lean leader is to constantly evaluate different aspects of the team in order to ensure that it is operating at peak efficiency. The leader will also need to keep up to date on customer requirements, as this is something that is going to be constantly changing as well. Doing so is one of the only truly reliable ways of staying ahead of the curve by making it possible to streamline the overall direction of the company towards the processes that will achieve the best results.

In order to ensure that this is the case, the Lean leader will want to make time in their schedule to look at the results and then compare them to the costs as a means of discovering the best ways to use all the resources available to them at the given time. This will include things like evaluating the organization as a whole in hopes of making it more efficient and reliable. It will also involve evaluating the value stream to ensure that it satisfies the customer on both the macro and micro level.

Focus on sustained improvement: It is also the task of the Lean leader to ensure that improvements that are undertaken are seen through to the end as well. This will often include teaching the team members the correct Lean behaviors to use in a given situation and also approaching instances of failure as opportunities for improvement and innovation.

Three actuals

Lean leaders typically use a different leadership style than many of their peers, largely because being a Lean leader requires an understanding that the best way to analyze a situation is to physically be in the space where the situation occurred. Once there, the Lean leader needs to consider what is known as the three actuals, the broadest of which is known as Genchi and is the issue that led the leader to come to the place in question. Genbutsu is the idea that it is important to view what is being created or provided in action before making any moves. Finally, Genjitsu says it is always best to gather as much information as possible before making a decision one way or the other.

Creating a Lean system

In order to create a Lean system that lasts, the first thing you will need to do is consider the absolute simplest means of getting your product or service out to the public and put that system into effect. From there, you will need to continuously monitor the processes you have put in place to support your business in order to ensure that improvement breakthroughs happen from time to time. The last step is to then implement any improvements as you come across. While there are plenty of theories and tools that can help you do go on from there, the fact of the matter is that creating a Lean system really is that simple. Many of the chapters in this book will consist of deep dives on various tools that will make this process as easy to set up and as most likely to stick as possible.

There's more to business than profits: When using the Lean system, the end goal is to determine the many ways that it might be possible to improve the efficiency of your business. While an increase in profits is often a natural result of this process, this should not be the primary motivating factor behind undertaking a Lean transformation. Instead, it is important to focus on streamlining as much as possible, regardless of what the upfront cost is going to be since you can confidentially expect every dollar you spend to come back to you in savings.

There are limits to this, of course, and at a certain point, the gains won't be worth the costs. To determine where this line is, you can use a simple value curve to determine how the changes will likely affect your bottom line. A value curve is often used to compare various products or services based on many relevant factors as well as the data on hand at the moment. In this instance, creating one to show the difference between a pre- and post-Lean state should make such decisions far easier to make.

Treat tools as what they are: When many new companies switch to a Lean style of doing things, they find it easy to slip into the trap of taking tools to the extreme, to the point that they follow them with near-religious fervor. It is important to keep in mind that the Lean principles are ultimately just guidelines and any Lean tools you use are just that, tools which are there to help your company work more effectively. This means that if they need to be tweaked to better serve your team and your customers, then there is nothing stopping you from doing just that. Your team should understand from the very beginning the limits and purpose of the Lean tools they are being provided and, most importantly, understand that they are not laws.

Prepare to follow through: Even if you bring in a trained professional to help your team over the initial Lean learning curve, it will still ultimately fall to you, as the team leader, to ensure that the learned practices don't fall by the wayside as time goes on. It takes time to take new ways of doing things and turn them into habits, and it will be your job to keep everyone on until everything clicks and they start operating via the new system without thinking about it. Likewise, it is important that you make it clear just why the Lean process is good for the team as a whole and for the individual team member as if they are personally invested in it, then it is far more likely that they will stick with it, even if the going gets tough.

Chapter 3: Setting Lean Goals

In order to eventually make the right changes to your business, the first thing you need to do is ensure that you set the right goals. In order to make sure that your goals will put you on the right track, you need to ensure they are SMART which means they are specific, measurable, attainable, realistic and timely.

Specific: Good goals are specific which means you want to be sure that the goal you choose is extremely clear, especially when you are first starting out, as goals that are less well-defined are much easier to avoid doing in favor of activities that provide more positive stimulation in a shorter period of time. Keeping specific goals in mind will instead make it much easier for you to go ahead and power through whatever task you are currently undertaking.

When you aren't quite sure if the goal you have chosen is specific enough to actually improve your chance of changing for the better, you may be able to figure it out by running through the who, why, where, when, and how of the goal. Specifically, you are going to want to consider who is going to be involved with you when it comes to the completion of the goal? What exactly is it that is going to be accomplished? Where will it be taking place, why it is important that you ensure it is completed as quickly as possible, and how exactly you can expect to go about doing it. Once you can answer all five of the big questions, then you know you have a goal that is specific enough to generate the type of results that you are looking for.

Measurable: SMART goals are those which can be broken down into small, easily manageable chunks that can be tackled one piece at a time. A measurable goal should make it easy to determine when exactly you are headed off track so that you can self-correct as quickly as possible. Measuring your progress will make it easier for you to keep up the good work.

Attainable: Perhaps more important than anything else, if a goal that you set is unattainable, especially the first goal that you set using this system, then you are going to unknowingly be wasting valuable time and energy while creating negative patterns that end in failure. What's more, you will end up reinforcing fixed mindset ideals, making this a bad choice any way that you look at it. This means that when it comes to setting goals, you are going to want to have a clear understanding of the current situation and anything going on with the business that would make it less likely to succeed as far as that goal is concerned.

Realistic: A good goal is one that is realistic, in addition to being attainable, which means that you can expect success without something extremely unlikely being required to push reality into your favor. An ideal goal is one that is going to require a good amount of work to achieve, while still remaining not too difficult as to become unrealistic. Additionally, you are going to want to shy away from goals that you can meet without putting any real amount of effort as goals that are too easy can actually be demotivating as it then becomes easy to continue putting them off until they eventually fade into oblivion.

Timely: Studies show that the human mind is more likely to actively engage in problem-solving behaviors when there is a time limit involved in the successful completion of the task in question. What this means for the goals you are setting is that if you have a firm completion date in mind for when you want to have reached your goal, then you will work harder in the period leading up to that date. This means that you are going to want to pick completion dates that are strict enough to truly motivate you to do whatever it is you have in mind, while at the same time, not being so strict that there is no realistic way that you can complete the task on time. The goal here is to throw a little extra hustle into your step and not force you to keep a grueling schedule, thus, ensure that you can always meet the schedule you set for the best results.

Policy Deployment

Also known by the name Hoshin Kanri, policy deployment is a way of ensuring any SMART goals that are set at the management level ultimately filter down to the rest of the team in a measurable way. Making proper use of policy deployment will essentially ensure that anything you are planning to put into effect doesn't accidentally end up creating more problems than it ultimately solves. It will also help to ensure as little waste as possible is generated as a result of things like inconsistent messaging from management or all around poor communication. The goal in this instance should not be to force various team members into acting in a specific way. It is about generating the type of vision for the business that everyone can appreciate and understand how it pertains to both the team and the customers.

Implementing the plan: Once all of the relevant SMART goals have been finalized, the next thing you will want to do is to group them together based on which members of the team will ultimately be tasked with solving them. Keep in mind that the fewer number of goals, the more likely it is that they will be acted upon in a reasonable timeframe. If your goals cannot be generalized in such a way, it is important to instead begin with the ones that are sure to make the biggest difference overall and work down the list from there.

Regardless of what goals you ultimately settle on, it is important that you take special care to ensure that there are no goals that do not have one person specifically assigned instructions to keep tabs on the overall progress while providing status reports when needed. This person should also be someone who can be counted on to make it clear to the other team members how important the goal is for the business as a whole and how it will make things easier in the long run.

Consider your tactics: Those who will be responsible for making the goal a reality should, in turn, be the ones who decide how the goal can best be completed by the team as a whole. However, this process should still include back and forth interaction between all levels of the team just to ensure that the tactics and the goal align properly. Tactics are likely to change as the goal heads towards success, which means it should be studied from time to time to ensure they remain appropriate for the goal in question.

Moving forward: Once the tactics have been agreed upon by all parties, it will then be time to actually put them into practice. This will be the stage where the team can really take over, though quality goals should still require buy-in from relevant parties. During this period it is important to ensure all communication from management is on message, to properly ensure that actions and broader goals will continue to align.

Review from time to time: It is important to keep in mind that once the action is in progress, the team leader will need to change the action as needed. This means that they will also be monitoring things as they proceed, hopefully, according to plan. Remember, Lean systems are always being improved upon, which means your goals and their implementation should be no different.

Chapter 4: Simplifying Lean as Much as Possible

All of the products and services that are generated by your business have a mixture of three different value streams that can ultimately be used productively if you take the time to understand them fully. These include the concept to launch stream, the creation of customer stream and the order to customer stream. In order to ensure you are getting the greatest overall value out of all of the processes your business finishes, it is important to look at a value stream map as it is an excellent way to ensure you are maximizing efficiency at every turn.

The average value stream map will include everything that ultimately comes together to generate value for the customer including activities, people, materials, and information. To properly visualize a value stream, you will want to follow the Plan/Do/Study/Act process, also known as the Lean cycle. To get started, you will want to plan out the task ahead by focusing on one goal at a time. From there, you will want to make a list of everything that will need to be done in order to ensure the task is finished successfully. This is then followed by the step of following through, the results being studied, and acted upon as required.

Create your own value stream map

A properly constructed value stream map is a vital part of the process as it will allow you to see the big picture by mapping out the entire flow of resources from their disparate starting points all the way through when they come together and eventually make it into the hands of the customer. As such, it then makes it far more of a manageable task to determine the points in the process that are bottlenecking the overall efficiency of your business's process and thus, taking the first steps towards adopting Lean processes.

While one person can certainly work through the following steps, the value stream maps that prove to be the most effective are often those created by the entire team, so that those who are the most knowledgeable about each step will be able to give their two cents as well. Your initial value stream map should be thought of as a very rough draft and should be constructed as such, which means planning it out in pencil and expecting lots of rewriting as you go along.

Consider the process: The first step in this process will be to consider exactly what it is you will be mapping. For businesses that are first starting out with the Lean system, you will want to begin by considering the various processes that are ultimately going to prove to be of the greatest value to the team as a whole and then work down the list from there. If you still can't decide where to start, then you will want to turn to your customers, consider what they have to say and start with the areas where you regularly receive the most complaints.

What is known as a pareto analysis is an effective tool at this juncture as it can make it easier for you to find the right place to start if you aren't sure where your efforts will be best put to use. It is a statistical analysis technique that can prove especially useful if you are looking at a few different tasks that are sure to generate serious results if only you could decide which one to use first. The goal, in this case, is to focus on the 20 percent of your business that, if nurtured, could ultimately generate 80 percent of your total results. Your initial value stream map may focus on only a single service or product or on multiples that share a significant portion of the process.

Choose your shorthand: The symbols you use to denote various stages of the process you are mapping don't really have any hard and firm guidelines as they are going to be unique to every project and every business. Regardless of what you and your team ultimately choose, it is important to create a list of all of the symbols you are using and what they mean so that anyone who comes in after the fact can easily get caught up. From there, it is important to stick to the designated symbols and not make anything up on the fly. Additionally, if the business is working on more than one value stream map at a time, it is important that the symbols correspond between the two. Otherwise, things can quickly spiral into illegibility.

Set limits: If taken from a broad enough scope, virtually every value stream map for your business can be connected to other value stream maps or go into greater detail. At some point, however, this is going to be counterproductive and you will have to set limits on what the value stream map is going to account for if you ever hope to successfully move forward. Likewise, if you let this part of the process get out of hand, then the map can lose focus and become less useful as a result.

Start with steps that are clearly defined: After you have a clear beginning and end for the process you are mapping, the next thing to do is to make a list of all of the logical steps that need to be taken from start to finish. This shouldn't be an in-depth look at every link in the chain, but instead, should be an overview of the major stages that will need to be looked at more closely as the process moves towards completion.

Consider the flow of information: One important step in the value stream mapping process that sets it apart from other similar mapping processes is that each value stream map also accounts for the way that information flows throughout the process from beginning to end. What's more, it will also chart the way information passes between team members as well. You will also need to ensure it takes into account the ways the customer interacts with your business, in addition to how frequently such interactions occur. You will also need to ensure the communication chain includes any suppliers or any other third parties the company deals with.

Further details: When it comes to breaking the process down to its most granular level, you may want to include a flow chart with your value stream map as well. A flow chart is a great way to map out the innermost details of how a given process reaches completion. This is also an excellent way to determine the types of muda you are dealing with, so you can consider if they can be removed from the process.

If you are interested in considering the ways your team physically moves around your space, then a string diagram can also prove effective. To generate this type of diagram, you will map out your business's workspace by drawing in what each member of the team has to do and where they have to go in order to fully complete the process. You will want to draw different team members or different teams in different colors to keep things from getting too confusing. From there, charting the flow of information as it relates to this data can lead to surprising conclusions regarding flaws that might otherwise go unnoticed for years.

Collecting data: When it comes to outlining your initial map of a value stream, you may find yourself with certain aspects of the process that require additional data before anything can be determined with any real degree of certainty. The data that you may need to track down will include:

- cycle time
- total inventory on hand
- availability of the service
- transition time
- uptime
- number of shifts required to complete the process
- total available working time

When it comes to collecting this data, it is important to always remember to go to the source directly and find the details you are looking for rather than making assumptions. Furthermore, it is important to get the most updated numbers possible as opposed to looking at older, more readily available figures or hypothetical benchmarks. This may mean something as hands-on as physically keeping an eye on every part of the process in question so you can take relevant notes.

Watch the inventory: Even if you are relatively certain about any inventory requirements for the process in question, it is vital that you double check before you commit anything to the value stream map. Minor miscalculations at this point could dramatically skew your overall results and essentially nullify all of your hard work if you aren't careful. This means you definitely need to adopt a measure twice in order to see the best results. After all, inventory is prone to building up for a wide variety of reasons and there is a good chance that you won't know it until you take a closer look and do a once over on what's really on hand. You can also use this step as an excuse to take stock of exactly what the team is working with and determine how far it will actually stretch effectively.

Using the data: After you have finished visualizing the steps found in your most important process, you are now ready to use it as a means of determining where any problem points might be. You will especially want to keep an eye out for processes that include redoing any previously completed work, anything that requires an extended period of resetting before work can begin again, or long gaps where parts of the team can do nothing except wait for someone else to finish, those that take up more resources than your research indicates you should or even just those that seem to take longer than they should for no particular reason.

Generate the ideal version of the value stream: After you have determined where the bottlenecks are occurring, you will want to create an updated value stream map that represents how you want the process to proceed once you have everything properly sorted. This will provide you with an A to C scenario, where figuring out the pain points represents B. Ideally, it will also provide a clue as to how you can go about eliminating the waste from the process in order to create an idea, which you can really strive for both in the short and the long-term.

Once you have determined the ideal state for the process, you can work out a future value stream map that will serve as a plan on how to take the team from where you are currently to where you need to be. This type of plan is often broken down into sections that last a few months, depending on what needs to be done. Additionally, most future value stream maps will come with multiple iterations because they will need to change several times as the project nears completion.

When working through various variations of the value stream map, it is important to pay close attention to the lead time available for various processes. The lead time is the amount of time it will take to complete a given task in the process and, if not utilized as efficiently as possible, it can easily lead to a wealth of bottlenecks. Remember, when it comes to creating the best value stream map possible, no part of the process is beyond scrutiny.

Chapter 5: Lean and Production

When putting together a Lean system for your business, you will eventually determine where the waste is hiding in your current processes, which is when it will be time to consider what can be done about the flow of the process. Often, the answer will come in either the form of a continuous flow model or a batch model.

Continuous flow model: The ideal version of the continuous flow model sees the customer order a product or service before the necessary steps are taken to generate the product or service that is being paid for. The product or service is then delivered to the customer who then pays for their order. The end result here is that there is no downtime between when the customer puts in their request and when it is completed. Furthermore, every step is going to smoothly flow into the next as a means of ensuring that overall downtime is reduced as much as possible.

The biggest upside to the continuous flow model is that it allows business to make assumptions and plan for the future based on a profit level that prioritizes continuity and stability. A continuous flow setup also features less waste than other types of processes. The biggest downside, however, is that this type of scenario can be difficult to produce reliably as every step in the process is rarely equal, regardless of how clear the value stream map might be. If you are striving to create a continuous process scenario, then you should be aware that new problems can also appear quickly if your available margin for error begins to shrink.

In order to persist despite these drawbacks, you will want to do your best to attack these problems head-on and be determined to push through them if you hope to find success. Additionally, if you hope to choose this route, it is important to start your journey to a Lean system with this in mind, as a continuous system is only going to work if every part of the system is completely in sync with all the rest.

Heijunka is a useful tool when it comes to facilitating this process as it promotes leveling out the quantity and quality of the process over a prolonged period of time in hopes of making everything as efficient as possible and, what's more, to expressly prevent batching. While it might sound complicated, in reality, this process can be as simple as making sure your team has all of the storage space they need to organize the various parts of the project. They store them in folders that are organized based both on the frequency of use as well as the due date. Folders that are currently in use can be stacked vertically on top of one another while those that are idle can be stored horizontally out of the way somewhere. There are also numerous other types of organizational methods that promote heijunka, so it can be helpful to explore them all to see what offers the most benefit to your business.

Batch production: Unlike with the continuous flow model, with the batch production model, the steps in the process to create the product or service are all completed in bulk one after the next. This makes it the superior choice when it comes to situations where what the process generates is evergreen as this will allow it to be stockpiled as a direct counter to erratic customer demand. Depending on the specifics of your business, batch production can also dramatically decrease the associated production costs as few team members can move from step to step instead of having all the steps operating at once. It also provides lots of opportunities when it comes to cross-training.

As a general rule, you can count on batch production to be less productive when there are a greater number of individual steps that are required to complete the process. This is because the starting and stopping times would need to be calculated for each which can add up quickly if batch sizes are quite large. This can also potentially create a delay if a customer places a large customized order when a batch is already in the middle of production.

Takt time

Short for the word Taktzeit, Takt time is a variation of the Japanese principle of measuring time, despite its German name. Despite the fact that it is primarily used in production environments, it can have a beneficial effect on most tasks performed in a business environment as well. Specifically, Takt time is the time it takes for a team to start a new process after completely finishing the last, assuming the production rate is equal to the rate of customer demand.

Determining takt time: If your team completes processes one at a time throughout the workday, the takt time of that process can then be determined by taking the time that has elapsed between two processes, assuming of course that demand is still being met. This means it can be written as $T = Ta/D$. In this case, T is your Takt time, Ta is the amount of time available to finish processes, and D is the amount of demand that the process experiences.

You will not want to automatically take these results as fact, however, as it is rare to find a team that can run at peak efficiency at all times. As such, when it comes to determining an accurate takt time, you will want to add in some wiggle room here to compensate for the fact. From there, you will want to adjust your takt time based on additional customer requirements or team demands.

Takt time benefits: Once you have determined the accurate takt time for your business's processes, you will find that a number of additional beneficial options open up to you. First and foremost, you will find that it is clear which steps in the process are the bottlenecks which will make it easier to take steps to mitigate them specifically. Likewise, if you have any processes that typically go off the rails, that problem will be made apparent as well.

As a general rule, takt time places additional emphasis on steps that add value to the process as a whole, which makes it easier to use if you are looking for muda to remove from the process. What's more, once the team gets used to the concept of takt time, they will find that it is much easier to track how productive they are being throughout the day.

Be aware: Takt time is not a set-it and forget-it type of affair which means that if you find that your demand changes dramatically, then you will need to recalibrate all of your takt time to adjust for this fact. This also means that if your demand isn't relatively stable, then determining your takt time might not be very beneficial on its own. If you try and force your process into a takt timetable and it isn't a good fit, then all you will end up doing is causing even more waste in the long run.

Likewise, you will need to be aware of the way in which the products or services produced by your processes fit together or else, you risk creating bottlenecks anyway which will throw off the accuracy of your takt time. As a general rule, the shorter the takt time, the greater the amount of strain that resources including both machinery and people will experience on a regular basis.

Chapter 6: Run a Lean Office

One of the truly great parts of the Lean system is the potential it holds when it comes to standardization, specifically when it comes to minimizing waste. Much like when it comes to setting goals, setting work parameters that are clearly standardized makes it easier to answer specific questions about the process. This should include things like who will follow through on the process once it has been outlined, how many people will it take, what will the end result be, what the metrics for success should be, what is required to meet them, how long the process will take and more. These are all questions that ultimately need to be asked in order to guarantee your standardization measures don't end up creating new problems instead of solving existing ones.

Workflow standardization is not expressly designed to ensure that processes are completed as quickly as possible. Rather, it is about utilizing the most effective practices possible in order to ensure they are completed with the same level of reliable quality each and every time. You will also do well to remember that standard practices will naturally change over time as improvements to safety, quality, and productivity are found. You will want to take care to avoid becoming so reliant on a single type of standardization that you end up actually allowing it to hold you back from future progress.

With that being said, it is also important to avoid falling into the trap of undertaking a round of standardization solely for standardization's sake. Instead, it is important to consider if standardization is really the right choice by considering the various processes already in place and asking yourself if they would be of a higher quality, performed to a higher safety standard or completed with less waste. If you move forward, and this is not the case, then all you are doing is inviting in waste.

Furthermore, standardization should involve more than simple instructional documents. It should be created from the input of those who perform the processes on the regular and then combined with a fresh round of customer feedback to ensure all bases are covered. The reasons for the standardization process should be clear to everyone involved before getting started for the best results.

KPIs

KPIs, also known as key performance indicators, are extremely useful when it comes to determining the ideal steps to take during the standardization process. KPIs are also useful when it comes to measuring the overall success of the process as a whole based on numerous different metrics. Choosing the right KPI to focus on is a matter of considering what metrics you value most at the moment as well as in the long-term. There are a wide variety of indicators to choose from, all of which are useful in different circumstances and when it comes to accurately defining specific values. Essentially, each KPI can be considered an object which is useful in various value-add scenarios.

Choosing indicators: When it comes to identifying the KPIs you want to use for your business, the first thing you will need to do is ensure that your process is already well-defined as this will help you have a true handle on the specifics of every aspect of the process as well as the best ways to determine the ideal means of completion. It is important to only stick with indicators that are relevant to the goal you are currently working towards. Otherwise, they can easily be altered dramatically by factors that are literally outside of your control.

Much like with your goals, it is important that your KPIs are SMART and that they clearly indicate specific information for a specific purpose. You will also want to choose options that are easily measured while still providing accurate results if at all possible. Much like goals, KPIs are useless if they are not realistically achievable. The most effective KPIs are those that are relevant to the success of the business in the moment or in the future while also including an element of time that has specific periods as they relate to the data.

Be aware: It is also important to keep in mind that while determining specific KPIs isn't too difficult, keeping track and compiling the relevant data can be more difficult than it first appears. Furthermore, additional values, including those for things such as staff morale, are difficult to gauge accurately. Before you invest resources into generating KPIs, it is important to first make sure they are adequately measurable and useful. Otherwise, you will be on your way to creating even more waste.

You will also need to ensure the focus remains on keeping the KPIs on the data that they are detailing and use it as a means of determining the overall health of the business as opposed to a set of numbers that can only move one way. If your team ends up too focused on reaching a predetermined KPI, the data they return will be biased and inaccurate.

Chapter 7: Kanban

Kanban is a method of scheduling that is often used once a Lean system has been put in place. It serves as a type of inventory management whose end goal is to minimize waste in the supply chain. It also tends to come in handy when it comes to pinpointing problems as it makes these problem areas stand out more than they otherwise would. You will also find it useful when it comes to locating the upper end of work related to inventory that is currently underway to ensure the process doesn't overload.

This is a demand-driven system which means, it is often implemented as a means of ensuring quicker turnaround times while at the same time limiting the required inventory and increasing the overall level of competitiveness between the implementation team. It was first put into effect by Toyota in the 1940s after the company performed a study on supermarkets and decided to use similar practices in order to keep their factories optimally stocked at all times. This eventually became one of the core Kanban ideas of keeping inventory amounts level with consumption rates. Additional supplies are then added based on a predetermined set of signals to ensure that stock remains near the ideal level at all times instead of dipping low or overflowing at certain points. The signals in question are all based upon customer signals which means they can change at the moment if needed.

Kanban rules

- Each process creates an amount set by the Kanban.
- Following processes collects the number of items set by the Kanban.

- Nothing is created or moved with a Kanban.
- Kanban is attached to related goods.
- Defective products are not counted in the Kanban.
- The fewer the Kanban, the more sensitive the system is.

Kanban cards: Kanban cards are the means by which signals are used to keep the entire team on track as it moves through the process. While they were actual cards when the system was created, these days, there is a wide variety of software out there that will provide the relevant signals without bringing physical cards into the process. Kanban cards generally represent consumption via a lack of cards in one area which, by necessity, drives another part of the process to do what needs to be done in order to pass the relevant cards along.

These days, the electronic Kanban system is even more effective than its physical predecessor, making it a sure thing to ensure that cards get where they need to be when they need to be there. These systems often mark set types of inventory with specific barcodes that are then scanned throughout the process. Each scan then sends a specific message to the Kanban program which routes it as needed.

Kanban types: There are two main varieties of Kanban systems namely production systems and transportation systems. Production systems are sent as a means of authorizing production or a specific number of items, while the transportation systems are used as a means of authorizing the movement of specific items once they have been created.

Three bin system: An example of a basic type of kanban system is the simple three bin system for the supplied parts in scenarios where manufacturing does not take place in-house. One bin represents the factory floor (or the primary point of demand anywhere else), the second bin represents the factory store (the control point for the inventory), and the final bin represents the supplier. The bins then can have removable cards containing relevant product details along with any other important information.

When the factory floor bin empties out because the relevant parts were all taken up by various parts of the manufacturing process, the empty bin, and thus its kanban card, are then returned to the factory store (also known as the inventory control point). The factory store then replaces the empty factory floor with the full factory store bin which also contains its own kanban card. The factory store then sends the empty bin and its related kanban card on to a supplier. This, in turn, causes the full product bin from the supplier to eventually replace the empty bin on the factory floor and the process starts all over again. Thus, the process never runs out of product. This could also be described as a closed loop, since it provides the exact amount required, with only one spare bin so there is never oversupply. This 'spare' bin allows for uncertainties in supply, use, and transport in the inventory system. A good kanban system calculates just enough kanban cards for each product. Most factories that use kanban use the colored board system

Chapter 8: 5s

When it comes to determining what wasteful processes you are dealing with, it is important to ensure the work environment is in optimum shape for the best results. The 5S organizational methodology is one commonly used system based around a number of Japanese words that, when taken together, are first rate when it comes to improving efficiency and effectiveness by clearly identifying and storing items in their designated space each and every time.

The goal here is to allow for standardization across a variety of processes which will ultimately generate significant time savings in the long-term. The reason it is so effective is that each time the human eye tracks across a messy workspace, it takes a fraction of a second to locate what it is looking for and process everything around it. While this might not be much if it happens now and then, if it is happening constantly across an entire team, then it can add up to serious time loss when taken across the sum total of the process in question

Sorting: Sorting is all about doing what can be done in order to always keep the workplace clean of anything that isn't required. When sorting, it is important to organize the space in such a way that it removes anything that would create an obstacle towards the completion of the task at hand. You will want to ensure that process-critical items all have a unique space that is labeled as well as a space that is designated for those things that simply don't fit anywhere else. Moving forward, this will make it easier to keep the space free of new distractions. Nevertheless, it will still be important to encourage team members to prune their personal space regularly to keep new obstacles from popping up.

Set in order: When it comes to organizing the items in the workspace themselves, it is important to ensure all the items are organized in the order that they will most likely be used. While doing so, it is important to take care to ensure that everything required for the most common steps remains readily at hand to reduce movement waste as much as possible. Over time, keeping things in the same place will ensure that the process can be completed faster each time as muscle memory takes over, and team members are able to reach for things without looking for them.

It is important to keep an open mind during this step since ensuring that the workspace is set up in such a way that ease of workflow is promoted may require more than a simple organization, it may require a serious rework of existing facilities. Additionally, ensuring everything is arranged correctly will make it easier for you to create steps for each part of the process that anyone new to that part of the process can follow.

Shine: Keeping the workspace clean is an essential part of maintaining the most effective workspace possible. It is important to emphasize the importance of daily cleaning both for the overall efficiency boost and its ability to ensure that everything is where it is supposed to be so that there are no issues the next time they are needed. This will also provide an opportunity to have a regular maintenance if any is needed, which will serve to make the office a safer place for everyone. The end goal should be that any member of the team should be able to enter a new space and understand where the key items are located in less than five minutes.

Standardize: The standardize step is all about making sure the organizational process itself is organized in such a way that it can be applied throughout the entire business structure. This will make it easier to maintain order when things get hectic and also ensure that everyone can be held to the same reliable standard.

Sustain: Sustaining the process is vital as taking a week or more to properly get everything in order only to have it all fall apart six months later is going to accomplish nothing in the long-term. As such, it is important to ensure that the organization is a vital part of the DNA of the business in moving forward. If things are truly sustainable in this regard, then team members will be able to successfully move through the process without expressly being asked to. Unfortunately, you won't be able to expect this type of sustainability overnight. It will require plenty of training and an adoption of the idea as part of the business's culture.

Great starter tool: If your plan for your business is to transition to additional advanced Lean concepts over time, then 5S is a great way to start moving employees in that direction. It is especially effective with employees who are extremely stuck in their ways as, once they initially get on board, they will be hard pressed to deny the benefits in completion times that come with the improved organizational version. This, in turn, will make it easier for them to get on board with additional changes that may come in the future.

As a rule, when rolling out a new system like this, you can expect team members to only care about two things, the way the new system is going to affect them specifically and if the Lean process has actually seen results. This is also what makes 5S a great starting point as it has easily understandable answers for each that anyone can understand once they see the first workspace transformed for efficiency.

Knowing if 5s is right for your business: While 5S is a great choice for some businesses, it is not a one-size-fits-all solution, which means it is important to understand both of its strengths and its weaknesses when moving forward. Perhaps its biggest strength is that when implemented successfully, it is sure to help your team define their processes more easily while also helping them claim more ownership of the processes they are associated with as well. This extra structure also has the potential to lead to a much greater degree of personal responsibility among team members which will lead to a greater feeling of accountability throughout the process. When everything goes according to plan, this will then lead to further improved performance and better working conditions for everyone involved.

What's more, implementing 5S also has the potential to more likely make long-term employee contributions thanks to an internalized sense of improvement. Ideally, this will continue until the idea of continuous improvement becomes the order of the day. When done correctly, using 5S will also provide further insight into the realm of value analysis, equipment reliability, and work standardization.

On the other hand, the biggest weakness of 5S is that if it, and its purpose, are not communicated properly, then team members can make the mistake of seeing it as the end goal and not a means to an end. 5S should be the flagbearer for success to come in the future, not the sum total of a company's journey into Lean processes. Specifically, businesses whose movement is constrained significantly by external factors will have a hard time using 5S, and companies that currently have a storage problem would do well to solve it before attempting a 5S transition.

Additionally, just because 5S is a great fit for many companies doesn't mean that it will be the best choice for your team. This is especially true for smaller teams or for teams where team members wear many hats. Just because it is a popular way to implement Lean principles doesn't mean that it is going to be right for everyone. Moving ahead anyway and enforcing organization simply for the sake of organization won't do much of anything when it comes to generating real results. Instead, it will only generate new waste and it will only continue to do so before it is abandoned entirely.

This is especially true for businesses that run on a wide variety of human interaction, various management styles, and other management tools. However, when the various aspects work together properly, they will actually end up generating extra value for the customer which is a vital part of any successful business. If you blindly press forward with a 5S mentality, however, then it can become easy to lose sight of the outcome for the customer in pursuit of a perfect outcome or a perfect implementation of 5S principles.

Above all else, when implementing 5S, it is important that you stress to your team that 5S is something that should be part of the natural work routine and standard best practices, not an additional task to be done outside of daily work. The goal of 5S is to enhance the effectiveness of the workflow at every step in the process. Separating out the 5S into its own separate layer is the complete opposite of what the process stands for.

Chapter 9: Six Sigma

Six Sigma is the shorthand name given to a system of measuring quality with a goal of getting as close to perfection as possible. A company operating in perfect synchronicity with Six Sigma would generate as few as 3.4 defects per million attempts at a given process. Zshift is the name given to the available deviations between a process that has been completed poorly and one that has been completed perfectly.

The standard Z-shift is one with a number of 4.5, while the ultimate value is a 6. Processes that have not been viewed through the Six Sigma lens typically earn around a 1.5.

Zshift Levels: A Six Sigma level of 1 means that your customers will get what they expect roughly 30 percent of the time. A Six Sigma level of 2 means that roughly 70 percent of the time, your customers will get what they expect. A Six Sigma level of 3 means that roughly 93 percent of the time, your customers will be satisfied. A Six Sigma level of 4 means that your customers will be satisfied more than 99 percent of the time. A Six Sigma level of 5 or 6 indicates a satisfaction percentage of even closer to 100 percent.

Six Sigma Certification Levels: Six Sigma is broken into numerous certification levels depending on the amount of knowledge the person in question has regarding the Six Sigma system. The executive level is made up of management team members who are in charge of actively setting up Six Sigma in your company. A Champion in Six Sigma is someone who can lead projects and be the voice of those projects specifically.

White belts are the rank-and-file workers; they have an understanding of Six Sigma, but it is limited. Yellow belts are active members on Six Sigma project teams who are allowed to determine improvements in some areas. Green belts are those who work with black belts on high-level projects while also running their own yellow belt projects. Black belts lead high-level projects while mentoring and supporting those at other tiers. Master black belts are those who are typically brought in specifically to implement Six Sigma and can mentor and teach anyone at any level.

Implementation: Giving your team a compelling reason to try Six Sigma is vital to the overall success of the process. In order to ensure that Six Sigma is properly implemented, it is important that you properly motivate your team by explaining how crucial the adoption of a new methodology really is. The most common choice in these situations is to create what is known as a burning platform scenario.

A burning platform is a motivational tactic wherein you explain that the situation the company now finds itself in is so dire (like standing on a burning platform) that only by implementing Six Sigma is there any chance of long-term survival for the company. Having stats that back up your assertions is helpful, though, if times aren't really so tough, a bit of exaggeration never hurt. Adapting to Six Sigma can be difficult, especially for older employees and a little external motivation can make the change more palatable.

Ensure the tools for self-improvement are readily available: Once the initial round of training regarding Six Sigma has been completed, it is important that you have a strong mentorship program in play while also making additional refresher materials readily available to those who need them. The worst thing that can happen at this point is for a team member who is confused about one of the finer points of Six Sigma to try and find additional answers only to be rebuffed due to lack of resources.

Not only will they walk away still confused, but they will also be rebuffed for trying and not rewarded for taking an interest in the subject matter. A team member who cannot easily find answers to their questions is a team member who will not follow Six Sigma processes when it really counts.

Key principles: Lean Six Sigma works based on the common acceptance of five laws. The first is the law of the market which means that the customer needs to be considered first before any decision is made. The second is the law of flexibility wherein the best processes are those that can be used for the greatest number of disparate functions. The third is the law of focus which states that a business should only focus on the problem the business is having as opposed to the business itself. The fourth is the law of velocity which says that the greater the number of steps in a process, the less efficient it is. Finally, the last is the law of complexity which says that simpler processes are always superior to more complicated ones.

Choosing the best process: When it comes to deciding what process to apply the Six Sigma treatment to, the best place to start is with any processes that are already defective and need work to reduce the number of times they occur. From there, it will simply be a matter of looking for instances where takt time is out of whack before looking into those steps where the number of available resources can be reduced as well.

Methodologies: There are two main ways to get the most out of Six Sigma, DMADV and DMAIC.

DMAIC is an acronym that is useful when it comes to remembering five phases that can be useful when it comes to creating new processes.

- Define what the process should do based on customer input.
- Measure the parameters that the process will adhere to and ensure it is being created properly by gathering relative information.
- Analyze the information you have gathered.
- Improve the process based on the analysis you have completed.
- Control the process as much as you can by finding ways to reliably decrease the appearance of delinquent variations.

DMADV, on the other hand, also has five phases that correspond to the DMAIC phases.

- Define the solutions the process should be providing.
- Measure the specifics of the process to determine its parameters.
- Analyze the data you have collected up to this point.
- Design the new process using your analysis.
- Verify the results as needed.

Deciding if Six Sigma is the right choice: While the Six Sigma system has something to offer teams of all shapes and sizes, that doesn't mean that it is going to be the best fit for all of them. This is especially true as implementing it successfully depends on numerous different specifics, starting with the conviction of those who are looking to implement the system in the first place as well as the company's overall culture. This is why it is best to start with something less high-impact like 5S as a way to ease your team into things that are more of an overall change like Six Sigma.

When deciding if a Six Sigma transition is feasible, it is important to ensure that it is not seen as a fad and can instead be seen as an evolution of the ideals already in place. Generally speaking, the more involved the team leadership is from the beginning, the more onboard the rest of the team will be as well. It is important that the company culture is perceived to be one that supports this sort of positive change and to remember that if the management team can't come to a consensus on the new program, then it is sure to be dead in the water. This doesn't mean that absolutely every member of the team needs to be committed to the idea of Six Sigma from the start, but it does mean that the change needs to be institutional so the public front always needs to appear united.

After all, Six Sigma was founded on the idea of leaders mentoring those beneath them in order to ensure Six Sigma works as it should, but this need to be a full-time job for some people, at least until the new habits start to solidify among the team as a whole. Once this occurs, you can then count on the team members to keep one another on track. To ensure they get to this point, you are going to want to let them know that their progress is being tracked so that every team member constantly feels as though they are improving up until the point where they internalize the Six Sigma principles.

Conclusion

Thank you for making it through to the end of *Lean Enterprise: The Complete Step-by-Step Startup Guide to Building a Lean Business Using Six Sigma, Kanban & 5s Methodologies*. Let's hope it was informative and able to provide you with all of the tools you need to achieve your Lean implementation goals. Just because you've finished this book doesn't mean there is nothing left to learn on the topic. Expanding your horizons is the only way to find the mastery you seek.

When it comes to implementing Lean techniques successfully, it is important to be realistic when it comes to the timeframe required to not just ensure the entire team is up to speed, but that they have internalized the core Lean principles you are trying to instill. You will need to take a long hard look at your team and your business as a whole and decide where the most work is going to need to take place. Every business has limited resources, after all. It is important to think wisely prior to allocating them.

While you can easily get sucked into a pattern of changing everything, in order to ensure your business really is as Lean as possible, you should keep in mind that discretion is the better part of valor and you should be sure to start by focusing on those things that will end up doing the most good before moving on from there. Don't forget, change for the sake of change won't do anyone any good and will likely serve to create more waste than it will eliminate. Ultimately, it is important to remember that creating a Lean business is a marathon, not a sprint, which means slow and steady wins the race.

Finally, if you found this book useful in any way, a review on Amazon is always appreciated!

Lean Analytics

The Complete Guide to Using Data to Track, Optimize and Build a Better and Faster Startup Business

Introduction

Congratulations on getting a copy of *Lean Analytics: The Complete Guide to Using Data to Track, Optimize and Build a Better and Faster Startup Business* and thank you for doing so.

The following chapters will cover everything you need to know to get started with the process of Lean Analytics. The Lean Support system is a great way to ensure that your business is as efficient as possible by eliminating the amount of waste that is present. The Lean Analytics section is going to help with data collection and analysis. Thus, you'll determine where the waste is present, and this will help you to pick the right metrics to implement.

This book will discuss Lean Analytics and how its processes can help you reduce waste and find the best strategy to improve your business. It is just one step in the Lean Support System, but it is an extremely critical step. This guidebook will provide you with the information that you need to get started so that you can become an expert in Lean Analytics in no time.

There are plenty of books on this subject on the market, so thanks again for choosing this one! Please enjoy!

Chapter 1: What is Lean Analytics?

The central idea behind Lean Analytics is on enabling a business to track and then optimize the metric that will matter the most to their initiative, project, or current product.

There is often a myriad of methods to improve your product, but you may not have the time to work on all of them. With Lean Analytics, you will learn how to find and address the one thing that will make the biggest difference.

Setting the goal of focusing on the right method will help you see real results. Just because your business has the ability and the tools to track many things at once, does not mean that it would be in your best interest to do so.

Tracking several types of data simultaneously can be a great waste of energy and resources and may distract you from the actual problems. Instead, you will want to focus your energy on determining that one vital metric. This metric will make the difference in the product or service that you provide.

The method in your search for this metric will vary depending on your field of business and several other factors. The way that you'll find this metric is through an in-depth understanding of two factors:

- The business or the project on which you're presently working.
- The stage of innovation that you are currently in.

Now that we have a basic understanding of Lean Analytics and what it means let's take some time to further explore and see its different parts.

What is Lean?

Lean is a method that is used to help improve a process or a product on a continuous basis. This works to eliminate the waste of energy and resources in all your endeavors. It is based on the idea of constant respect for people and your customers, as well as the goal of continuously working on incremental improvements to better your business.

Lean is a methodology that is vast and covers many aspects of business. This guidebook will spend sufficient time discussing a specific part of Lean, Lean Analytics. Here, you can learn how to make the right changes. Of course, you will need a working understanding of where to start, and Lean Analytics can help.

Lean is a method that was originally implemented for manufacturing. The idea was to try to eliminate wastes of all kinds in a business, allowing them to provide great customer service and a great product while increasing profits at the same time. Despite its beginnings, the Lean methodology has expanded to work in almost any kind of business. As long as you provide a product or a service to a customer, you can use the Lean methodology to help improve efficiency and profits.

For instance, how will you determine which metric will help you succeed? Which metric will prove to be the best and result in the most improvement compared to others? How will the metric help, how should it be implemented, and how can you ascertain if it's successful in the end? Lean Analytics can help you gather the necessary information to find and work with the right metric.

Lean Analytics

Lean Analytics is part of the methodology for a lean startup, and it consists of three elements: building, measuring, and learning. These elements are going to form up a Lean Analytics Cycle of product development, which will quickly build up to an MVP, or Minimum Viable Product. When done properly, it can help you to make smart decisions provided you use the measurements that are accurate with Lean Analytics.

Remember, Lean Analytics is just a part of the Lean startup methodology. Thus, it will only cover a part of the entire Lean methodology. Specifically, Lean Analytics will focus on the part of the cycle that discusses measurements and learning.

It is never a good idea to just jump in and hope that things turn out well for you. The Lean methodology is all about experimenting and finding out exactly what your customers want. This helps you to feel confident that you are providing your customers with a product you know they want. Lean Analytics is an important step to ensuring that you get all the information you need to make these important decisions.

Before your company decides to apply this methodology, you must clearly know what you need to track, why you are tracking it, and the techniques you are using to track it.

Focus on the fundamentals

There are several principles of Lean that you will need to focus on when you work with Lean Analytics. These include:

- A strive for perfection
- A system for pull through
- Maintain the flow of the business
- Work to improve the value stream by purging all types of waste
- Respect and engage the people or the customers
- Focus on delivering as much value to the customer as effectively as possible

Waste and the Lean System

One of the most significant things that you will be addressing with Lean Analytics, or with any of the other parts of the Lean methodology, is waste. Waste is going to cost a company time and money and often frustrates the customer in the process. Whether it is because of product construction, defects, overproduction, or poor customer service, it ends up harming the company's bottom line.

There are several different types of waste that you will address when working with the Lean system. The most common types that you will encounter with your Lean Analytics include:

- **Logistics:** Take a look at the way the business handles the transportation of the service or product. You can see if there are is unnecessary movement of information, materials, or parts in the different sections of the process. These unnecessary steps and movements can end up costing your business a lot of money, especially if they are repeated on a regular basis. This will help you see if more efficient methods exist.

- **Waiting:** Are facilities, systems, parts, or people idle? Do people spend much of their time without tasks despite the availability of work or do facilities stay empty? Inefficient conditions can cost the business a lot of money while each part waits for the work cycle to finish. You want to make sure that your workers are taking the optimal steps to get the work done, without having to waste time and energy.

- **Overproduction:** Here, you'll need to take a look at customer demand and determine whether production matches this demand or is in excess. Check if the creation of the product is faster or in a larger quantity than the customer's demand. Any time that you make more products than the customer needs, you are going to run into trouble with spending too much on those products. As a business, you need to learn what your customer wants and needs, so you make just the amount that you can sell.

- **Defects:** Determine the parts of the process that may result in an unacceptable product or service for the customer. If defects do exist, decide whether you should refocus to ensure that money is not lost.

- **Inventory:** Take a look at the entire inventory, including both finished and unfinished products. Check for any pending work, raw materials, or finished goods that are not being used and do not have value to them.

- **Movement:** You can also look to see if there is any wasted movement, particularly with goods, equipment, people, and materials. If there is, can you find ways to reduce this waste to help save money?

- **Extra processing:** Look into any existing extra work, and how much is performed beyond the standard that is required by the customer. Extra processing can ensure that you are not putting in any more time and money than what is needed.

How Lean can help you define and then improve a value stream

Any time that you look at the value stream, you will see all the information, people, materials, and activities that need to flow and cooperate to provide value to your customers. You need these to come together well so that the customer gets the value they expect, and at the time and way, they want it. Identifying the value stream will be possible by using a value stream map.

You can improve your value stream with the Plan-Do-Check-Act process. This strategy can be used upfront so that you can design the right processes and products before they reach their finished form. Additionally, the strategy helps you to create an environment that is safe and orderly and allows easy detection of any waste.

Another method of creating this environment is the 5S+ (Five S plus): sort, straighten, scrub, systematize, and standardize. Afterward, ensure that any unsafe conditions along the way are eliminated.

The reason that you will want to do the sorting and cleaning is to make it easier to detect any waste. When everything is a mess, and everyone is having trouble figuring out what goes where, sorting and cleaning can address waste quite fast. There will also be times when you deem something as waste and then find out that it is actually important.

When everything is straightened out, you can make more sense of the processes in front of you. Afterward, you can take some time to look deeper into the system and eliminate anything that might be considered as waste or unsafe, and spend your time and money on parts of the process that actually provide value for your customer.

Chapter 2: The Lean Analytic Stages Each Company Needs to Follow

To be successful with Lean Analytics, you'll need to follow several different stages. You won't be able to move on to the next stage if you do not complete the preceding step. There are five in particular that you will need to focus on to get work done with this section of the Lean support methodology. The five stages are:

- **Stage 1:** The initial stage is where you will concentrate on finding the problem for which people are searching for a solution. A business that focuses on business to business selling is going to find this stage critical. When you address this problem, then you can move on to the next stage.

- **Stage 2:** For this stage, you are going to create an MVP product that can be used by early adopter customers. This stage is where you are aiming for user retention and engagement, and you can spend some time learning how this will happen when people start to use the product. You can also learn this information based on how the customer uses your site and how long they stay. You'll take some time at this stage because you will need to experiment and also may need to go through and choose from a few different products before you get the one that is right for you. Once you have this information, you can move on.

- **Stage 3:** Once you find out how the early adopter customers are going to respond to a product or service, it is time to find the most cost-efficient way to reach more customers. Once you have a plan ready to get those customers, and then more of them start purchasing the product, then you can move to the next stage. You would not want to go with a product that may be popular but costs a ton of money, which will cut into your revenues and can make it difficult to keep growing in the future.

- **Stage 4:** You are now going to spend some time on economics and focusing on how much revenue you are making. You want to be able to optimize the revenue, so you need to calculate out the LTV:CAC ratio. LTV is the revenue that you expect to get from the customer, and the CAC is the cost that you incurred to acquire that customer. You can find this ratio by dividing your LTV by the CAC. Your margins are doing well if you get an LTV that is three times higher than the CAC. The higher the margins you get, the better because that means you are going to earn more in profits from the endeavor.

- **Stage 5:** In the final stage, you will then take actions that are necessary to grow the business. You can continue with your current plan if you are making a high enough margin from the previous steps, or you may need to make some changes to ensure that you can earn enough revenue to keep the business growing. You can also spend time making plans on where you would like to concentrate on in the future to increase the growth of your business and help it expand. The main goal for your business is to keep growing and increase revenue. This step helps you to reevaluate what you have in your current plan and decide if it is working for you or if you need to go with a different option.

Chapter 3: The Lean Analytics Cycle

The Lean Analytics Cycle is vital in helping you get started on this part of the Lean support methodology with your business. There are four steps that will come with this process, and following each one can be crucial in ensuring that this works for you.

The best way to think about the Lean Analytics Cycle is like the scientific method. You need to do some thinking to determine what needs to be improved in your business, form a hypothesis to help lead your findings, and then perform experiments to see if that is the right process for you to keep following. If things don't work out, you don't just give up. You will continue to find new experiments, going with the same hypothesis if it works (otherwise you'll need to form a new hypothesis) until you find the right solution.

The Lean Analytics Cycle will be incredibly helpful when you begin going through the entire process. Let's take a look at the steps that you need to fulfill to use the Lean Analytics Cycle.

What do I need to improve?

Before you can do anything with the Lean Analytics Cycle, you must really understand your business. You need to know all the important aspects of your business, in addition to knowing what you want to change.

During this first step, you may need to talk to other businessmen to help you find what metric you should use, based on what is most relevant to your business right now. You may also want to take a look at your business model to find out what metric will work best for you.

After you have time to choose a metric, you should connect it to the KPI or the Key Performance Indicator. An example of this is the metric that is seen as a conversion rate if the KPI is the number of people who currently purchase the product.

To make this step easier, the first thing that you would want to do is write down three metrics that are important for your business. Afterward, write down the KPI that would be measured for each metric.

Never try to implement the Lean system without understanding the most important processes that need to be improved. Sure, you could probably make a long list of things that you may want to improve in your business. But you won't really see the benefits of the Lean system if you don't pick things that are important to the overall functioning of your business. Look closely at what your business needs to improve, and pick the one that is the most important before moving on.

Form a hypothesis

This is a stage where a level of creativity needs to come into play. The hypothesis is going to give you the answers that you need to move forward. You will need to look for inspiration, and you can find it in one of two ways. You can look for an answer for something like "If I perform ____, I believe ____ will happen, and ____ will be the outcome."

The first place you can look into is any data that you have available. Often, this data will provide you with the answer that you need. If you do not have data at all, you may need to do some studying of your own to come up with an answer. You could use some of the strategies from your competitors, follow the practices that have worked well for others, do a survey, or study the market to see what the best option will be.

What you need to keep in mind here is that the hypothesis is there to help you to think like your audience. You want to keep asking questions until you understand what they are thinking, or learn to understand the behavior of your audience or customer.

Conduct an experiment

After you have taken the time to form a hypothesis, it is time to test it out with the help of an experiment. There are three questions that you need to consider to get started with an experiment:

- **Who is the target audience?** You need to carefully consider who your customers are and whether or not they are the right customers, or if you should look somewhere else to get better results. Also, think about some of the ways that you reach them, and if there are better ways to do this.

- **What do you expect the target audience to do?** This often includes purchasing the product, using the product, or something similar. You can then figure out if the audience understands what you want them to do; is it easy for them to do this action, and how many of the target audience are completing the task?

- **Why do you think they should accomplish the action**? Are you providing them with the right motivation to accomplish the task? Do you think that the strategy is working? If they aren't being motivated enough by you, are they doing these things for the competitors or otherwise?

Answering these questions is vital because they may help you understand your customer better than ever before. Creating your experiment during this stage does not have to be difficult. Try using the following sentence to help you get started:

"WHO will do WHAT because WHY to improve your KPI towards the defined goals or target."

If you have gone through and come up with a good hypothesis in the previous step, then it shouldn't be too hard to create a good experiment as well. Then, once you have the experiment, you can go through and set up the Lean Analytics so that you can measure your KPI and carry on in the experiment.

Measure your outcomes and make a decision

You can't just get started with an experiment and then walk away from it. You need to measure how well it goes to determine if it is truly working; if some changes are needed; or if you need to work from scratch. You can then make a decision on the next steps you need to take. Some of the things to look for when measuring the outcomes during this stage include:

- **Was the experiment a success?** If it is, then the metric is done. You can move on to finding the next metric to help your business.

- **Did the experiment fail?** Then it is time to revise the hypothesis. You should stop and take some time to figure out why the experiment failed so that you have a better chance at a good hypothesis the next time.

- **The experiment moved but was not close to the defined goal.** In this scenario, you will still need to define brand new experiment. You can stay with the hypothesis if it still seems viable, but you would need to change up the experiment.

Chapter 4: False Metrics vs. Meaningful Metrics

One of the prerequisites for working with Lean Analytics is understanding that most people are using their data wrong. When you don't use your data correctly, you are not going to be able to come up with patterns, opportunities, or results that are achievable.

There are two points that come with this idea. These two points are:

- There are many companies, as well as people, who will label themselves with descriptions like "data-driven." Sure, they may use up a lot of their resources on compiling data. However, they then miss out on the "driven" part. Few are actually going to base a strategy on the information that they acquire from the data. They may have the right data, but they either don't understand it or choose to react to it incorrectly.

- Even if the actions of a company or person are driven by data, the problem of using wrong data still exists. Often, they will oversimplify these metrics and then use them according to the convention. Keep this in mind: just because other people do this or have done this doesn't mean it is going to prove useful to the goals that you have. Consequently, the data is going to become garbage in, and then the analysis is garbage out. This is often known as false data.

As a business who is interested in working with Lean Analytics, it is important to learn the difference between false metrics and meaningful metrics. If you follow false metrics, you are going to be following a strategy that is not going to help you reach your goals, which will mean a lot of time, effort, and resources wasted.

The biggest false metrics to watch out for

As a business that is trying to cut out waste and ensure that you provide the best customer service and the best products possible, you must always ensure that you watch out for some of the false metrics that may come up. Many people who don't understand how data works will be taken in by these false metrics that, in reality, will mean wasted time and resources. Some of the most common false metrics for you to watch out for include:

- **The number of hits:** Just because you have a website that is attractive and contains many points of interest, doesn't necessarily mean that it will tell you what the users are really interested in. You should not focus on the number of hits your website gets. This may make you feel good about your website, and it can be neat to see how many people come and visit your website. But you need to focus more on what the customer is interested in or looking for.

- **Page views:** This metric refers to how many pages are clicked on a site during a given time. This is a slightly better than hits, but you typically don't want to waste your time with this metric. In most cases, unless you are working with a business that does depend on page views, such as advertising, the better metric for you to use is to count people. You can do this with tools that will provide information on unique visitors per month.

- **Number of visitors:** The biggest problem with this metric is that it is often too broad. Does this type of metric talk about one person who visited the same site a hundred times, or a hundred people who visited once? You most likely want to look at the second group of people because you've obtained more impressions. Otherwise, just looking at the number of visitors will not give you this information.

- **Number of unique visitors:** This is a metric that is going to tell you how many people got to your website and saw the home page. This may sound good at first, but it is not going to give you any valuable information. You may also want to find out things like how many visitors left right away when they saw the page or how many stayed and looked around. Unique visitors can help you see that some new people are coming onto your website and checking things out but they don't really tell you much about those visitors or what they are doing.

- **Number of likes, followers, or friends**: This is a good example of a vanity metric that is going to show you some false popularity. A better metric that you can go with is the level of influence that you have. What this means is how many people who will do what you want them to do. While it is good to have followers and likes on your page to show that people are looking at your content, it is not as important as some of the other metrics that you can pick.

- **Email addresses:** Having a big list of email addresses is not a bad thing by itself. But just because you have this large list does not mean that everyone on it is going to open, read, and act on the messages that you send out. You want to make sure that the email addresses that you do have are high quality and are from people who actually want to hear from you, even if that means your email list is a little bit smaller. If you are collecting emails, strive to get addresses from people who are actually interested in your product and services. Don't just collect emails so you can boast of a large list.

- **The number of downloads:** This is a common metric that is used for downloadable products. While it can help with your rankings in the marketplace when you are in the app store, the download number is not going to tell you anything in depth, and it won't give you any real value. If you would like to get some precise answers here, you can pick some better metrics. The Launch Rate is a good place to start because it will show the percentage of those who downloaded, created an account, and then used the product. You can also use something like Percentage of Users Who Pay so you can see how many actually pay for anything.

- **Time spent by customers on a page or website:** The only time that this is going to be useful is for businesses that are tied directly to the behavior of the engaged time. For example, a customer could spend a lot of time on your web page, but what if they are spending that time on the help pages or on the complaints pages? This is not necessarily a good thing for your business, so this metric is not the best one for you to go with.

If you want to pick out a metric that will actually help your business get ahead, then you must make sure that you avoid some of these false metrics. They may look good on the surface, but in reality, they are just giving you information that could be pretty useless, and you will end up wasting a lot of time and money to follow them.

Chapter 5: Recognizing and Choosing a Good Metric

Part of the Lean Analytics methodology is finding a good metric to help you out. The Lean Analytics Cycle is a measurement of movement towards a goal that you already defined. So, once you have taken the time to define your business goals, then you must also think about the measurements you can make to progress towards the goals.

This can be hard to do. How are you supposed to find a good metric that can make sure you go towards the goals that you set out? Some of the characteristics that you can look for when searching for a good metric include:

- **Comparable:** You know that you have a metric that is good if it is comparable. You want to be able to compare how things have changed in the last year, or even from one month to another. This gives you a good idea if there have been any changes, positive or negative, with your business process, customer satisfaction, and more. You can ask yourself these questions about the metric to help test for this:
    - How was the metric last year, or even last month?
    - Is the rate of conversion increasing? You can use the Cohort Analysis to help with conversion rate tracking.
- **Understandable:** The metric that you use should never be complex or complicated. Everyone should be able to understand what it is. This ensures that they know what the metric is measuring.

- **Ratio:** You should never work with absolute numbers when you are working with metrics. If you find that you have those, you should try to convert them to make comparisons easier, which in turn makes it easier to make decisions.

- **Adaptability:** If you have chosen a good metric, it should change the way that the business changes. If you notice that the metric is moving, but you have no idea why it is moving, then it is never a good metric. The metric should move with you, not randomly on its own, or it won't be a secure one to use.

Types of Metrics

There are two metric types that you are able to use when doing Lean Analytics. These include qualitative and quantitative metrics. To start, qualitative means that the metric has a direct contact with your customers. This would be things such as feedback and interviews. It is going to provide you with some detailed knowledge of the metric.

You can also work with quantitative metrics. These are more of a number form of metrics. You can use these to ask the right types of questions from the customer.

Of course, both of these methods have other things under them that make them easier to use. You will find that both of these methods have actionable and vanity metrics.

- **Vanity metrics** will not end up changing the behavior of the thing you are concerned about. These are a big waste of your time, and you should avoid them as much as possible. They seem to provide you with some good advice and something that you can act upon, but often they don't lead you anywhere and can make things more difficult. If you are working with a company to help you determine your metrics, be very wary if they start touting the benefits of following any of the vanity metrics.

- **Actionable metrics** are going to end up changing the behavior of the thing you are concerned about. These are the types of metrics that you want to work with on your project. They are metrics that can lead you to the plan that you should follow and can make it easier to come up with a strategy to make your business more efficient.

- **Reporting metrics** is a good way to find out how well the business is performing when it does even everyday activities.

- **Exploratory metrics** are going to be useful for helping you to find out any facts that you do not know about the business.

- **Lagging metrics** are good to work with when you want more of a history of the organization and you want as many details as possible to help with a decision. The churn of a company can be a good example of the lagging metrics. This is because it is going to show you how many customers have canceled their orders for a specific amount of time.

- **Leading metrics** are good because they can help provide you with the information that you need to make future forecasts for the business. Customer complaints can be a good example of leading metrics because it can help you to predict how a customer will react.

You will need to determine which kind of metric you want to use based on the problem or project that you are working on. Working with one metric is usually best. Doing so will help keep you on track, so you know what to look for. Don't waste your time trying to work on more than one metric. You will only get confused and end up with no clear idea about the strategy to follow.

Chapter 6: Simple Analytical Tests to Use

Another thing that you should concentrate on to do well with Lean Analytics is to have some familiarity with the tests that are used. These tests are helpful because they are going to be used to help you examine any assumptions that you are trying to use here. These tools can also be used to help you identify customer feedback so you can respond to them properly. Let us take a look at some of the best analytical tests that you can use when working with Lean Analytics.

Segmentation

The first test is segmentation. This process involves comparing a set of data from a demographic bucket. You can divide up the demographics in any manner that you choose such as gender, lifestyle, age, or where they live. You can use this information to find out where people are purchasing a particular product; if there are different buying behaviors between female and male customers; and if your target audience seems to be in a certain age group or not.

The reason that you want to build up a user segment is to make it easier for the data to be actionable. Analytics can teach you a ton about the people who purchase from you, but there is often a lot of information there, and it can be hard to draw good conclusions from this information. After all, while this information from the past can be useful, it isn't going to be the best to tell you how to improve either retention or conversion rate.

This is where the process of segmentation is going to come into play. When you learn how to filter out the audience, you will then be better able to create a plan to make new products that serve them the most. Analytics can give you the information that you need, but segmentation can help you to act.

For example, you may have a conversion rate that seems average or good, but it could be from a combination of one group that converts really high and consistently so, and then another group that seems to never convert at all. You could be wasting a lot of money on that second group where you are hardly getting anybody to convert at all. Segmentation can be used to help you understand what things you are doing the right way when engaging the first group, and can give you a plan on how you can improve to work on that second group.

With segmentation, you don't want to only look at the data to learn some more about your users, but you also want to come up with data that you can act upon. Segmentation can help you with this. You will be able to divide up the people in your customer base and learn how to advertise to them better than ever before.

Remember that not all customers are going to be the same. There are some of your customers who may purchase something once, and they aren't regular customers. While it is still good to reach out to them, you want to learn who your regular audience is, what they respond to, and what keeps them coming back. This is going to ensure that you keep them coming back and earn as much profit as possible.

So, how do you create a segment of your customers? There are many different options that you can use when creating a segmentation. But let's look at the process that you can use to create a segmentation for your Lean Analytics project. The steps you need to use include:

- **Define the purpose of your segmentation:** You should first figure out how you want to use your segmentation. Do you want to use it to get more customers? Do you want to use it to manage a portfolio for your current customers? Do you want to reduce waste, become more efficient, or something else? Defining your purpose can make it easier to know how you should segment out your customers.

- **Identify the variables are the most critical:** These are going to help influence the purpose of your segmentation. Make sure that you list them out in order of their importance, and you can use options like a Decision tree or Clustering to help. For example, if you want to do a segmentation of products to find out which ones are the most profitable, you would have parameters that are revenue and cost.

- Once you have your variables, you will need to **identify the threshold and granularity of creating these segments.** These should have about two to three levels with each variable identified. But sometimes you will need to adapt based on the complexity of the problem you are trying to solve.

- **Assign customers to each of the cells.** You can then see if there is a fair distribution for them. If you don't see this, you can look for the reasons why, or you can tweak the thresholds to make it work. You can perform these steps again until you get a distribution that is fair.

- **Include this new segmentation in the analysis** and then take some time to look it over at the segment level.

Cohort Analysis

The Cohort Analysis is a test involves comparing sets of data using a time bucket. In this test, there will be differences in behavior between customers who arrived at the free trial stage of your process, versus those who showed up at the initial launch, and then those who are in the full payment stage.

Each of these is significant because it helps you to figure out which customers are likely to come back and be full-fledged customers when in the future. Those that show up in the initial stages when the product is free are often not the customers you are going to see when sales actually start. They may have just wanted to try it out and didn't really have an investment in the product.

Those that are in the later two stages can be customers who are better for you to work with. They will be the most interested in the product because they invested some money to get it. You really want to study these using the cohort analysis to figure out who your real customer base is and how they behave so that you can better market to them later on.

A/B Tests

A/B testing is a process where you examine an attribute between two choices. This could be something like an image, slogan, or color so that you can figure out which option is the most effective choice.

Let's say that you had two products that you are comparing and you want to find out which ones customers liked the best. Did they choose one product over the other and why? Did they respond better to the choice that was in green or the one in blue?

For this test to really work, you must assume that everything else is going to stay the same. So, it would have to be the same product, but there is one variable that is different between them. You could put up a website, for example, and have a red background on one version and a yellow background on another. Then you could use A/B testing to figure out which one the customer responded to the best out of those choices.

In addition, you can also work on multivariate analysis. This is pretty much the same thing, but instead of going through and testing out one attribute, you will go through and compare several changes against another group of changes to see which is the most effective. This one will require there to be a few changes in the second product compared to the first to be the most effective.

There are several keys that you need to have in place when you are ready to do an A/B test. These include:

- Know the reason that you are running this A/B test.

- The item that you are testing needs to be noticeable to the audience. If you make a minor change that no one is going to notice, then your results are not going to be that reliable.

- Stick with testing just one variable at a time. If you go through and do multivariate testing, or test more than one thing at a time, you will run into trouble. You may not know for sure which variable is causing the changes you see.

- Your test needs to end up being statistically significant. This means that it must have a sample size that is big enough to test and know that the results are valid within a certain margin of error.

Let's take a look at an example of how to do this. We are going to use this test on a website that you are trying to improve. There are two main ways that you can do this including:

- You will test the pages on separate pages.

- You will use JavaScript to conduct the test inside the page, so you don't need two different URLs to do it.

The first option is going to mean that you will need to have two different URLs for the pages that you are testing out. You can make them similar names, but make sure that there is some way that you can keep track of them and not get the two confused.

With the second option, you will need to have some experience working with JavaScript. You can then place some of this code on the website so that it can dynamically serve one option or the other.

The method that you choose is often going to depend on the one that you like the most and which tools you want to use. Both of these will give you some valid results, but you will find that implementing each of them takes a different amount of time to set it up.

Chapter 7: Step 1 of the Lean Analytical Process: Understanding Your Project Type

Now that we have taken a look at some of the different part of Lean Analytics, it is time to take a closer look at how the process works. These can help you to get started with the Lean Analytics stage for your business and ensure that you are getting the most out of Lean.

The first step that we are going to look at is understanding your business or your project type. How are you supposed to pick out the right metrics if you have no idea what kind of business or project type you are working on? You must really understand the project at hand so that you can choose a fantastic metric that can show you results.

There are six general business types that you can fit into, and they all will have metrics that are going to work best or matter the most, for each one. If you see that your business or project is on this list, your job will be simple. You just need to focus your attention towards understanding the priorities of what needs to be measured. This can include in-depth external research.

However, if you have a business that is not on this list, this doesn't mean you are out of luck and can't do anything. You can just use some of the information that is in this chapter as an example and build up your own understanding and metrics from this chapter.

E-commerce

The first type of business is going to be e-commerce. These are growing like crazy right now as many customers are looking for the things they want to buy online more and more. And many companies find that they can make large profits by offering their products and services online to these customers. An e-commerce business is going to be any that has their customers buy from a web-based store. This could include businesses such as Expedia.com, Walmart.com, and more.

The strategy for this type of business is that you need to understand the customer relationship that you want. This means that you are going to focus either on new customer acquisition or customer loyalty? You have to decide between these two because this is going to help with all other decisions that you make with this type of business.

There are many metrics that you can choose to go with in an e-commerce business. Some of the typical ones that other companies have chosen in this industry include:

- Inventory availability
- Shipping time
- Mailing list and how effective it is
- Virality
- Search effectiveness
- Shopping cart abandonment
- Revenue that you make on each customer
- The amount you spend to get new customers

- Shopping cart size
- Repeat purchase
- Conversion rate

The best metrics

Of course, there are several metrics that will work the best and will provide you with the best return on investment, when working with an e-commerce site. The best metrics to use here include:

- **Conversion rate:** This is the percent of all visitors to your site who also purchase something. The average conversion when it comes to online retail is 2%. There are some that can do better though. For example, Tickets.com is over 11%, and Amazon.com is at almost 10%.
- **Shopping cart abandonment:** It is typical that 65% of the shoppers to your website are going to abandon their carts. Many of these are because of the high costs of shipping, and others are from the high price of all the items in their cart. You should definitely take some time to analyze any shopping cart abandonment that is happening in your business so that you can learn why you are losing these customers.
- **Search effectiveness:** The majority of your buyers are going to have to search to find what they need. If you make your search more effective, it can help your customers find what they want, rather than having them leave in frustration. Remember that about 79% of your total shoppers will use the search engine for half of the goods they want.

Software as a service

These types of companies are going to sell software in downloadable form or as a subscription. This can be things such as Skype, Evernote, Basecamp, Adobe, and more. They are not selling a physical product to someone, but these software programs are still pretty important for most people to get work done or to do other things on their computers.

The strategy with this one is that most software is going to consist of products that are on a subscription which means that retaining the customers is important. Your success is going to really depend on building up a loyal base of customers faster than those customers disappear.

There are some metrics that you can use to make this happen. Some of the most common metrics that are used with this type of business model include:

- Reliability and uptime
- Upselling
- Virality
- Customer churn
- Customer lifetime value
- Cost of getting new customers
- The amount of profit you make per customer
- User conversion
- User stickiness
- User enrollment
- User attention

The best metrics

Just like last time, you are able to use any of the metrics that are above, but there are some that could be the best for helping you reach your overall goals. Some of the best metrics to use with a software company includes:

- **Paid vs. free enrollment:** You will find that your enrollment rate is going to change depending on whether or not you asked for credit card information in the free stage or not. The former is going to get an average signup rate of 2 percent, and then 50 percent often end up buying. When you do not ask, the average may increase to ten percent, but only 25 percent purchase the product.

- **Growing revenues and upselling:** Some of the best software providers are able to get 2 percent of their paying subscribers to increase what they pay each month. Being able to grow your customer revenue by 20 percent in a year can be achieved if you work towards it.

- **Churn or attrition rate:** This is the percent of your customers who are leaving. Going across the industry, the top companies usually have an attrition rate between 1.5 and 3 percent each month. If you have a percentage that is higher, then you need to find ways to make the customers stay.

Mobile app companies

These are companies that are going to provide apps to be used on mobile devices like Android and iPhone. Some of the companies that can fall under this category would be ones like WhatsApp and Instagram.

The strategy that you want to go with here is to find the right target audience. There are a lot of ways for your app to make money, but you will find that the majority of your revenue is going to come from a smaller group of customers, rather than from the population as a whole. You should focus your analysis as well as the metrics you use to help you better understand those customers.

There are many metrics that you are able to use as an app company. Some of the most common options include:

- Customer lifetime value
- Churn rate
- Ratings click-through
- Virality
- The revenue you make from each paying user
- The revenue you make for each user
- Percentage of users who end up paying
- How much it costs to get the customers
- Launch rate
- Downloads

Best metrics

Of course, there are many metrics that you can choose to look at when it comes to being an app company, but a few of them are going to provide you with the most information and can help your business to really grow. Some of the best metrics you can use include:

- **Downloads and the app launches:** The number of people who download the product and then activate it will fit in here. It is known that quite a few people who decide to download an app will then never activate it or use it at all, especially if the app is free.

- **The cost to get new customers:** You can follow a general rule to have a budget of 75 cents per user in your marketing initiatives to help attract new customers. You should always make sure that the cost to get new customers is lower than what you will earn on them. So, if you will only earn 50 cents on a customer, then you shouldn't spend 75 cents on each one.

- **The average revenue you earn per customer:** This is often going to be determined through the business model. For example, Freemium apps, or apps that you receive revenue from engagement in the app, will often have a higher revenue per user compared to those that are premium apps.

Media site companies

If you are in this industry, you have a website that is going to provide some information, such as articles, in return for earning advertising or any other type of revenue. These would include most blogs and other sites like CNN.com, CNET, and more.

Media sites need to really understand the source of their

revenue. It is not coming directly from their readers or the people who use their "product," but it is coming from advertisers who are trying to reach those readers. So, if you are a media site company, you would get revenue from affiliates, click-based advertising, display advertising, and sponsorship. You would want to design your key metrics to work for this.

Some of the different metrics that you can choose to work with for a media site company include:

- Page inventory
- Pages per visit
- New visitors
- Unique visitors
- Content and advertising balance (you don't want too much advertising on the page, or it takes away from the content and keeps the customer away).
- Click through rates
- Ad rates
- Ad inventory
- Audience and churn

Best metrics

- **Click through rate:** This is the number of users that are going to click on a link out of all the users who check out the page. The average click-through rate for a paid search in 2010 is 2 percent, but some companies can get

higher. If you see that you are at one percent, then it is time to make some changes. But if you are above that number, you are doing really well.
- **Engaged time:** This is how long your reader will stay on the site and look through the content and the ads. Most media sites are going to aim for 90 seconds for content pages, and a little less with landing pages. If you find that your visitor is not spending more than a minute on the content pages, then it is likely your content is not engaging them.
- **Content optimization for media:** This one means taking the content that you already have and changing it so that it works on other venues, such as podcasts and video. You should track how others are using the materials you have because this can help you find some new opportunities to use.

User-generated content business

If you have a community that is engaged, they are going to contribute free content. And this same engagement is going to provide you with ads as well as other revenue sources. Some examples of companies that work with this include forums, Wikipedia.com, Reddit.com, Facebook.com, and Yelp.com.

The strategy that you should use is one that takes into account user engagement. This business is going to be successful when its visitors become regular contributes, and they interact with others in the community and provide quality content. User engagement tiers to measure involvement can be good as well.

Some of the different metrics that you may want to use with user-generated content include:

- Notification and mail effectiveness
- Content sharing
- Value of the content that is created
- Content creation
- Engagement funnel changes
- How many engaged visitors you have

The best metrics

- **Time on the site each day**: Here you are going to measure how long the average user is on your site and engaged on a typical day. This is a good thing to measure for engagement and stickiness. The average number is about 17 minutes a day, though Facebook is usually an hour, and Tumblr and Reddit are 21 and 17 minutes respectively.
- **Spam/Bad Content:** With these kinds of communities, you need to make sure that good content is always uploaded. You will have to spend time and money to keep bad content and fraudulent content off the site. You can measure what you think is good and bad and then build up a system to help keep up with this. You can also spend your time watching out for quality decline and then fix it before it ends up ruining your community.

Two-sided marketplace business

These kinds of businesses are going to connect buyers and sellers, and they will earn a commission on the work. It is kind of a variation of the e-commerce store. Some options of this would include Priceline.com, Airbnb.com, Ebay.com, and Etsy.com.

The strategy with this business is that you need to be able to attract in two different customers, the buyers and the sellers. The best bet is to focus on those that have the money to spend first. If you can find a group of people who want to spend their money, then those who want to make money will pretty much line up to do it.

Some of the metrics that you are able to use when it comes to a two-sided marketplace business include:

- The volume of sales and the revenue you earn
- Pricing metrics
- Ratings and any signs of fraud showing up
- Conversion Funnels
- Search effectiveness
- Inventory growth
- Buyer and seller growth

Best metrics

- **Transaction side:** Sellers usually won't have the money or time to analyze pricing and the effectiveness of their copy and pictures. As the owner, you will have the aggregate data from all your sellers, and you can use this information to help them with this analysis.

Transaction size is the same as the purchase size, and of course, it is going to differ based on your business type. You should help your sellers measure it so that they can understand the behavior of your buyer and use it to sell more items.

- **Top 10 lists:** You can make top ten lists to help your buyers find the best products, and your sellers to know what is going on in the industry and what they can do to be more profitable.

As you can see, there are many different types of businesses out there. And it is likely that your business is going to fit somewhere in this list. If it does, then there is an outline that you can use for developing a good strategy. Even if you don't, you can combine a few of these strategies to help you come up with the metrics, and the plan, that you need to succeed.

Chapter 8: Step 2: Determine Your Current State

Now that you know which business type you are in, it is time to move on to the second step of Lean Analytics. This one is going to require you to determine which innovation stage you are in right now. The one metric that means the most to you right now is a function of time. It is going to change as your project keeps moving on through the different stages of innovation. There are several different stages of innovation that you can work with. These include:

Stage 1: Empathy or is this a real problem?

In this stage, you are going to identify a problem in your business and then get inside the head of your potential user. You should be in their shoes and understand why there is a problem and what they are thinking. You may need to spend some time talking to potential customers to help with this stage. The more that you are able to talk to your potential customers and others in the market about the product or service you want to offer, the better off you will be. This can give you some real insights that can drive your business forward.

You need to focus on any metrics that are going to help you to determine whether or not the problem is harmful to your business. The metric needs to also determine if there are enough people who care about this problem. If only a few people see it as a problem, then it probably isn't worth your time taking care of it. But if a big percentage sees this as a problem, then it is something to take care of. You can also use metrics that will see what the success rates of your existing solutions are and if you need to change some of them.

Stage 2: Stickiness or do I have a good solution?

In this stage, you are going to start by making a Lean prototype of your solution to the problem you found in the previous step. You have to ask yourself whether or not people will pay for this. This is when you can gain feedback from small focus groups and testers. Based on that information, you can make adjustments and changes to the solution until you get it right.

You are going to need to focus on any metric that proves your solution will encourage the user to engage and also come back to your business.

Stage 3: Virality or does this solution provide value to enough people?

Once you have a solution and a product and they are seen as effective, you need to decide whether its value adds enough that the customer will tell others about it. Remember that word of mouth endorsements are valuable as a precursor to growth measurement and as free advertisement for the business. You can work for endorsements that are either natural (the customer enjoys the product or service enough that they just give out recommendations to their friends) or ones that are incentive-based (such as giving the product for free or at a discount).

For this one, you need to focus on metrics that are able to measure out if you are getting any new customers from your existing ones. And you want to know how many of these referrals are happening. You can also take a look at metrics that can check for how long it takes for news to spread or the cycle time.

Stage 4: Revenue or can I make this profitable?

Now you need to work on how much revenue you can expect from selling the product or service that you created in the last step. You can work on prices, standardization, control costs, and margins. You need to take the time in this step to prove that you are able to make money in a self-sustaining and scalable way, or this is not the solution for you.

This one is going to need you to focus on metrics that can tell you the net revenue that you are able to earn for each customer. The net is going to be the revenue that you make per customer minus the amount you spent to get that customer.

Stage 5: Scale or can we expand to a bigger audience?

Now that you have a product, you showed that it is effective, and you have a business model in place to show that it is going to be functional and profitable, you can now invest and expand it into new markets. This can include new geographies, channels, and audiences as well.

If you are dealing with a project or business that is oriented on efficiency, you need to focus on metrics that are able to reduce costs. If you are working in a business or project that is differentiation oriented, you will want to focus on metrics that will track margins for you.

What can I do with these innovation stages?

Now that you know a little bit more about these innovation stages, it is time to figure out where you are and learn what you can do with each one. The steps that you should take from here include:

- Look through the stages above and determine where your business or project is right now.

- Refocus on the things that you should be measuring at the stage you are in.

- If you find that your project does not fit into this framework, then it is important to remember that all innovative endeavors are going to follow a pattern of stages as well as maturity through time. Are you able to borrow this framework and leverage it in some manner so that you can figure out what stage your project is in right now? This can really make a difference in helping you to understand what you really need to be focusing on right now.

Knowing where you are in the innovative stage can make a big difference. When you look at the five stages above, you will have a clear outline of what you need to focus on and what needs to be done to keep you moving forward. If you have no idea where you are right now, then how are you supposed to know what steps to take to get to the next level? Always have a good idea of where you are in the innovation process, and then you have a clear picture of where you should go next and can keep on track.

Chapter 9: Step 3: Pinpoint the Most Pressing Metric

To help you to be successful with any type of innovative project, the key is to focus. Consequently, you can't spend your time on too many metrics because this is going to make you feel distracted and you are going to lose all your focus.

In the first few stages of innovation, it is often best to reduce the number of metrics that you track. If you can, you should focus on just one metric, the one that matters the most right at the moment. The metric that is the highest-priority is usually related to the most important project or business need.

For example, a subscription software company may be in the virality stage, and they are trying to gain traction with it. They may decide that the net adds metric is the one that will help them out the most. Remember that Net Adds = Total of New Paid Subscribers – Total That Cancelled.

There may be other metrics that your company can use, but you need to just focus on one. You will need to figure out what problem is the most pressing or the most important right now, and then go with a metric that fits with this the most.

How can I find that one metric?

Some of the steps that you can use to find the one metric that matters the most right now include:

- Write down the top three to five metrics that you really like, and you often track.

- How many of these metrics are actually any good and help you out?

- How many do you use to make business decisions? How many of those would actually be vanity or false metrics?

- What stage are you at with the business or the project? Do you really understand what matters in the business model? Can you discard any of the metrics that aren't really adding value to you right now?

- Are there any other metrics that are not on your list that you can think of and that you think could be more useful right now?

- Once you have written down all those metrics, you can go through the list and cross off any bad or false metrics. Add any new good ones that you think of on the bottom.

- Now that you have a list, you should go through and pick the one metric that you absolutely can't live without to help you with the project in its current stage.

What to do after optimizing that one metric

Once you have the metric and the project at a level where you are happy with the numbers, you must remember that you will need to continue measuring them. You never know when that project or that metric will need to be changed up again to help you in the future. But you can rest assured knowing that the process is now controlled and optimized. What this means is that you are now at a point where you are achieving a certain level of results.

Now you are able to go back to your list and work on the metric that is the next highest priority. This is going to be the next highest priority of your business, or the next biggest project need. You can review through the innovation stage you are currently at and the business or project type, and then you can determine which point of interest you should focus on.

Remember that the goal here is to only work on one metric at a time. There may be several metrics that need to be addressed in your business. Keep your focus on one at a time.

Sure, you can go through and write out a list of the different things that need to be addressed at some point, and Lean Analytics is a good time for this because it can help you see what problems are there. But you should pick the one that is the most pressing and work on that one first.

After you have time to complete the Lean Analytics stage on this problem and you have a winning strategy in place for it, then you can move on to the next step of picking out a new project to work on. You can implement this process on as many projects as you would like. Just make sure that you are only working on one at a time.

Chapter 10: Tips to Make Lean Analytics More Successful for You

Getting started with Lean Analytics is something that can take some time to get used to. It is going to provide you with great results and a winning strategy that is sure to get you ahead. But for those who are just starting out with this stage, or who are just getting started with the whole idea of the Lean system, you may need some help to get going on the right foot. Here are some great tips that you can follow to ensure that you are doing well with Lean Analytics and to ensure it is as successful as possible for you:

- **If you are doing an A/B test, you need a lot of users:** You are not going to get any good results from you A/B test if you don't have a lot of users to help you out. This means that it is not going to work all that well if you are a small startup or if there are not a lot of people you can measure. You should have a minimum of 10,000 events before you attempt this kind of test. These events can include visits or people who use a feature. Make sure that you are able to get this many users to help you out before you get started.

- **Make big changes:** If you are not able to see the changes from a few feet away, it is likely that the people you are testing with the A/B test won't either. For example, A/B tested 41 different shades of blue. The results were not the best because there were just too many different shades and for most people, they looked too similar. You need to make big changes before you do an A/B test, or it won't work well for you.

- **Measure the tests properly:** You are not going to get the right results if you are not properly measuring the tests. You need to have the right metric in place. Also, make sure that you never stop a running test too early, or you may miss out on some of the important results that you need.

- **Use the tools that you need:** Lean Analytics has a ton of tools that you can use to make it successful. Make sure that you are properly trained to handle each part and that you don't miss out on some important tools that can make this more successful.

- **Know where your business is now**: How are you supposed to have any idea of what kind of project to work on and what metrics to use if you don't have a good understanding of your business? Make sure that you know the overall goals and vision of the business. This can help you to spot some of the problems that you need to fix and can make it easier to ensure that whatever changes you do decide to make are going to go along with what your business is all about.

- **Understand the different metrics:** You should spend some time looking at the different metrics that are available for you to use on your project. Each one can be great, but it does depend on the type of business that you are running and the project that you want to work with. You need to learn which metric is going to be the right one for you.

- **Add this into the Lean Support System:** Many people are fond of the Lean Support System. This allows them to get rid of a lot of the waste that their company may have, and can make them more efficient. But you need to do Lean Analytics first to see success. This helps you to gather the information that is needed and then sort through it and analyze it. Then you can use this information to come up with the best plan to handle your problem. If you just jump into a strategy without the resource, it is likely that you won't see results at all.

- **Focus on the main problem first:** If you are like many businesses, there are probably many problems that you need to solve. But you don't have the time and resources to do all of them at the same time. When you get started with Lean Analytics, you must figure out what the main issue is, the one that will have the biggest impact on your profits, and work with that first. Once you have successfully implemented Lean Analytics and worked on the problem, then you can go back and see if there are any other problems that need to be addressed.

- **Get rid of the waste:** Remember that the most important thing that you will do with the Lean system is get rid of waste. And the data that you collect in the Lean Analytics stage is meant to help you to find the waste and learn how to get rid of it. Take a look at some of the most common types of waste that businesses may experience (and that we listed in an earlier chapter) to give you a good idea of where to start.

Lean Analytics can be a great way for you to get a strategy together that will help your business become more successful. If you follow these tips and some of the strategies that we talk about in this guidebook, you are sure to see some amazing results in no time.

Conclusion

Thank you for making it through to the end of *Lean Analytics: The Complete Guide to Using Data to Track, Optimize and Build a Better and Faster Startup Business*, let's hope it was informative and able to provide you with all of the tools you need to achieve your goals.

The next step is to start the process of implementing Lean Analytics into your own business. Learning how to make changes so that you can be more cost effective and provide better service to your customers all starts with Lean Analytics. This stage asks you to search for the data you need and analyze it so you know what step to take next. You can't come up with a plan for improving your business without the help of Lean Analytics to make it possible.

Finally, if you found this book useful in any way, a review on Amazon is always appreciated!

Agile Project
Management

The Complete Step-By-Step Beginner's Guide to Agile Project Management and Software Development

Introduction

Congratulations on downloading *Agile Project Management* and thank you for doing so.

Your project went completely off the rails despite putting forward your best efforts. Deadlines were missed, the scope was never attained, and fingers were pointed. No amount of efforts and dedication by your developers could make the situation better, and the client grew tired of asking for a refining of the product.

Answers were sought because popular phrases such as "We are never going to do this again" and "This should not have happened" went round in the conference rooms.

If you have ever gone through this, you are not alone. If you know how a Waterfall model works, then you can understand the problems that come with it. In brief, a Waterfall model is a sequential design where development assumes a downward flow like the shape of a waterfall.

Therefore, Agile Project Management provides solutions to many problems of the traditional waterfall model. The Agile methodology responds to the actual behavior of humans after every sprint. It eliminates the need to wait for years before releasing a final product to the client. Agile propels for fast release and quick response to user changes.

With Agile Project Management, one is sure of a faster development and product release. In this book, all the chapters help you learn Agile Project Management. It offers the reader with unique insights to help adopt Agile principles in the organization. It helps one change their priority from generating money for the company by satisfying the customer.

There are plenty of books on this subject on the market, so thanks again for choosing this one! Every effort was made to ensure it is full of as much useful information as possible. Please enjoy!

Chapter 1: What Do We Mean by Agile Project Management

Agile Project Management refers to a way of managing projects. In short, it is a way of handling a project to help fulfill its goals. This may include early delivery of a specific task and a continuous development of the project processes and products. Besides that, Agile Project Management focuses on creating a flexible scope and ensuring that products are tested to reflect the different needs of a customer.

Everyone understands that a project requires total attention, time, and correct planning. Without these factors, it becomes difficult for a project to be successful. All software projects have a goal and objective set. In addition, each project has defined time for completion.

Agile Project Management has different techniques that help improve the way of managing a project. Being a leader of a project, it is important to stay updated with the Agile approaches of project management. It is also useful if one can understand some history of the project management and the common issues that may be involved.

The Start of Modern Project Management

Projects have been around right from the invention of the printing press to the building of the Great Wall of China. Projects have been present whether small or big.

However, project management started in the middle of the twentieth century. It was during this time when most researchers were looking for major developments and changes in the computing industry. In order for these researchers to complete these developments, they decided to set up effective ways that can help them manage and complete projects. At the start, these mechanisms were based on a systematic procedure, so people in the field of programming and computing went on to adopt most of these processes.

One reason for this is because all computers relied mostly on hardware. The software was set to expand. In fact, during this time, software represented a tiny part of the computer. Don't forget that part of the history of computing was comprised of computers built with thousands of physical tubes and only a few lines of code. Therefore, the manufacturing process used during this period led to the development of the Waterfall methodology.

Why go Agile?

Now that you have some basic understanding of agile, it's better to also know how agile works, the roles of agile, and why adopt agile.

At the start, we had defined Agile Project Management as a way of managing projects by delivering value to customers and the organization. In other words, this gives project managers the ability to offer a high-priority and high-quality work in the various projects.

Agile Project Management is a dynamic approach to managing projects where it accommodates whichever type of change that comes on the way. Even late in the development phase, Agile Project Management will embrace the change. It will allow one to create the right features that have the best value. With Agile Project Management, one is sure to deal with real-time information as well as handle scope, time, and cost.

The most interesting thing about managing projects using Agile practices is that it is simple and efficient. It will help cut down on the complexity by ensuring that it reduces the time needed to put together requirements for the entire project. It will further help build the whole project and test it to discover multiple product problems.

Why do you need to bother about Agile Project Management?

Let's say you have started to apply Agile practices, and you find it much complex, chances are that certain things aren't going well. Maybe some components are missing. In this case, it is advised to inspect the implementation process. What everyone must know is that implementing Agile in the right way will always create success in the delivery process. In general, one must record a positive improvement both in the value delivered to customers and the product.

With Agile Project Management, time and cost are an important aspect. It continuously examines these two elements that are the keys to any software project. In addition, it delivers a rapid feedback to the team. It further helps the team to adapt and apply a QA practice. This will help deliver top-notch service and output. Project managers who have mastered Agile practices dwell on real-time delivery, proactive, and the aggregate flow. The overall goal is to have a minimum cost of a project as well as a working product delivered on time.

Examples of Agile Project Management Software

1. Monday

Monday is formerly called DaPulse. It mainly focuses on social communication and sharing of internal information. Monday has prioritized top choices in Agile project management. Collaboration is achieved by a board outlining who is working on what within a specific time. Then teams can move on to comment on other teammate's work or add required files. There is a mobile or desktop application that facilitates real-time notification. Monday is great software to use for basic and enterprise plans.

2. Wrike

It is SaaS project management and collaboration software. Wrike is designed based on a minimalist user interface. It has a project management feature that allows one to monitor dates, project dependencies, and manage assignments and resources. It features an interactive Gantt chart, sortable table and workload view that one can customize to store project data. The collaboration features of Wrike help in conversations, decision making, and asset creation by the team members. These comprise of Wrike's Live co-editor, tools to attach documents, track changes, and discussion threads. Wrike has an "inbox" feature and browser notification to remind users of updates from their dashboards about pending tasks. It is available both in IOS apps and native Android.

3. Asana

This is web-based software designed to enhance collaboration. It mainly allows users to control projects and online tasks without using an email. Asana supports team sharing, organization, planning, and monitoring progress of each member in a simple style.

4. Taiga

This is an open source Agile Project Management platform designed for smaller teams of project managers, developers, and designers. It facilitates project collaboration, time tracking, and task management. Taiga features a customizable Agile functionality like Kanban boards and backlogs. This software supports Web-based deployments that are flexible with a lot of operating systems. The system can be accessed as a free self-hosted model where projects are public. There is also a paid plan where projects are private.

5. Planbox

This supports members with different business functions to collaborate, plan, and create Agile projects. It has a Scrum methodology that features iterations, Scrum roles, backlog, sprints, and story points. Planbox has a four-level platform that includes tasks, projects, initiatives, and items. This supports drag and drop prioritization, to-do lists, messaging, bug tracking, and reporting among other functions.

6. Smartsheet

This is another SaaS-based application that supports collaboration and work management. Smartsheet has an interface similar to that of a spreadsheet to track projects, manage calendars, monitor progress, and manage other works. Every row in the Smartsheet might have files attached to it, discussion board linked to it as well as emails stored. While information is updated, another Smartsheet which monitor the same task is updated automatically.

7. Trello

It is web-based project management software and a well-known brand out there. With a free account, anyone has an opportunity to use the majority of the functions while a premium account has complex features. Trello has a design that depends on the Kanban methodology. Projects are viewed as boards that have lists. Each list has a progressive card that supports drag-and-drop functionality.

Chapter 2: How to Implement Agile Project Management

The previous chapter defined Agile Project Management. If you remember, Agile Project Management represents an iterative approach that takes into consideration the user feedback, adapts to changes, and produces a working result.

This chapter shall look at how one can implement Agile Project Management. Project managers who want to learn how they can use Agile practices in their organization can benefit more from this chapter.

Let's first start by looking at Agile Project Management in detail:

• Agile is an iterative practice. This means that it is applied in small chunks. In each chunk, there is an improvement based on the previous feedback.

• Agile is defined as an approach and mindset. It is not a collection of instruction. In fact, it's a mistake to consider Agile as a black and white template.

• Agile delivers a practical working result after every iteration process. This involves creating a rough draft and revising the draft based on the feedback from the client.

• Agile Project Management involves communicating effectively over a chain of emails or meetings. It involves communicating proficiently in the right way and in a precise manner.

Agile practices provide fast development, more revenue, and more releases. So, why should you not use Agile Project Management for the organization or team? Well, when it comes to using Agile to building projects, you don't just throw in your documentation, tools, and plans. Even though they are important, the main things that one should consider are the iteration, collaboration, and prototypes.

Most users don't focus a lot on documentation or in the long-term plan. Instead, whatever they want, they want it delivered immediately. A majority would want a problem fixed at that time instead of waiting for one month.

These days, users have a lot of needs, and there is no better way than using Agile to satisfy these needs.

How can you tell whether Agile is the best for your team?

Agile is an amazing way to build projects, but not every project can deliver all the benefits that Agile practices provide.

Agile redefines the working process of an organization. It changes everything right from the time when a team of developers begins to work on a project to the time when the project requirements were outlined. It reduces the time it takes to build a project. Therefore, all project managers must understand if their organization can handle all the changes that Agile Project Management brings in the project cycle. To be sure, below are a few questions to consider.

1. **Are you ready to begin a project without having a precise endpoint?**
   Agile Project Management involves completing small

parts of a project within a short period. There is real-time testing of a product with users as well as a recording of feedbacks about the product. If a person is not used to this type of process, it can be stressful and tiresome. That is why, it is important that everyone in the team are prepared and comfortable with releasing a half-baked product to test with users before the organization finally decides if they should adopt the Agile development practices.

2. **What are your levels of risk-taking?**

A major characteristic of Agile projects involves a continuous release of a product and learning from the mistakes highlighted by the users. This could be a high-risk endeavor for those people who are not used to taking risks. The ability to take a big risk and begin an organizational change is a major decision. So, before fully adopting Agile practices, it is important to be ready to deal with any unfamiliar challenges that might come on the way.

3. **Measure the level of flexibility of your team**

When it comes to building Agile products, one has to work directly with customers to improve the product. It is not similar to how developers and designers create products based on how they think a product should be. Instead, the product is created based on the customer's feedback.

4. **How is the discipline in the organization hierarchy?**

One of the key factors with Agile is not just working with users alone but the link between developers and the major stakeholders. There are some companies where this is not easy. In fact, some have a complicated hierarchy where it is hard to get in touch with the stakeholders. Ask yourself, what kind of hierarchy exists in your company?

5. **What are the means you use to measure success and progress?**

Agile features a continuous work to refine and make a product better. Therefore, an individual who is quick to adopt a new idea and abandon the previous one might not get the best results that Agile provides. It is important to take some time and redefine the measures of success or progress to help achieve the goals set.

Implementation of Agile

Project managers have a responsibility to ensure that everybody works in the correct way. They have to make sure that a team stays put on the requirements that a client has stated. With Agile, project managers have an easy time to oversee a project. Below are seven steps in the implementation of Agile in an organization.

Step 1: Define a vision

What is it?

At the start of an Agile project, it is required that a clear vision of the project is defined. This means the project manager and the team sit down to determine the reference ground of the project. For a product company, the way to know the vision of a project is to use an Elevator Pitch. However, companies that are building something different other than a product, they need to modify the Pitch and make strategies that suit the needs of the company.

Who should be present?

In this stage, one has to find out the prospective clients of the product. It is advised to invite major stakeholders such as managers, product owners, and company directors.

When should it happen?

It is important for the strategy meeting to take place before the start of any project to help keep the mission of the project alive.

How long should it last?

While this depends on you, a good strategy meeting should last between 4-16 hours, but it should have some breaks in between.

Step 2: Create a roadmap for your product

What is it?

Once the strategy is approved, the product owner should transform that strategy into a product roadmap. This is a complicated view of the project requirements that have a loose timeframe for when to create each project.

The 'loose' timeframe is very critical here because you can't spend a whole month planning for each step. Therefore, one should aim to prioritize, identify, and create a rough estimate of each product piece. This will facilitate the delivery of a product that is ready to be used.

Things that should be present

It is the role of the product owner to create the product roadmap, but it should also have some buy-in and contribution from other stakeholders who take part in the project.

When should it take place?

Make sure that a roadmap exists before planning for the sprints. Therefore, it is better to begin building a product once the strategy meeting is over.

How long should that last?

Agile Project Management requires rapid development of a product. There should be no delay in the early stages of planning. However, the product roadmap is the mission of the product, and it has to last until that time when the product owner can feel that everything has been covered.

Step 3: Get ready with a released plan

What is it?

So far, there is already a strategy in place and a plan that defines the time. At this point, it is the role of the product owner to create an advanced schedule that will define the subsequent dates of release of the working software. Keep in mind that Agile projects have numerous product release before the final product was released. Therefore, a product owner has the option to place emphasis on the product features he or she wants to launch first.

For example, if a project started on November, one can schedule to release it on February before again releasing it in May when the most advanced features have been developed. However, this depends on the level of complexity of the project and the length of the sprints.

Who should be present?

The top members of the organization should be present on the day of product release. Team members and managers should be present, as well. A few stakeholders will also help get the team fired up.

When should it happen?

It is good to ensure that the release of a working product takes place on the scheduled date.

How long should that take?

Be realistic about the length of a session. Pay attention that it does not slow things down. An interval of 4-8 hours should be appropriate.

Step 4: Plan your sprints

What is it?

This should be the time to change from a macro view to the micro view. Make sure that the sprints are short with specific things to accomplish. You can set the sprint to last for 1-4 weeks and maintain the same length for the whole project since this will help teams to plan in an accurate manner.

When starting a sprint cycle, make sure that a team builds a list which includes a backlog of items to complete in a given timeframe. This will help one to release functional software.

Who needs to be present?

This is a collective responsibility of the whole team. In other words, the project manager, the owner of the product, and the team members should be present to speak out their suggestions about the product.

When should it take place?

Planning for a sprint always takes place at the start of every sprint cycle.

How long should it last?

Remember that by planning for a sprint, it sets the tone for the cycle, although, you don't want to take a lot of time at this stage. It should last for 2-4 hours. Also, once the sprints are planned, the team can start.

Step 5: Ensure your team is on track

In the entire process of creating sprints, there must be time to deal with all the challenges that come on the way. A lot of these challenges are those that can slow down the time to achieve the goals. Therefore, a project manager should organize for daily meetings with their team. The meetings can last for 15 minutes.

Some of the things to discuss in the meeting include:

- What did the team complete on the previous day?
- Current tasks that the team is working on.
- Challenges that the team has experienced.

Although this may sound like a waste of time to some of the members of the team, holding such kinds of meeting help clarify certain things in the meeting. Remember that Agile practices call for a quick response to problems, and raising these problems in public is the best way to develop team collaboration.

Step 6: You have done sprints. Time to review

What is it?

Let's assume that everything has gone well according to plan. Therefore, before the end of each sprint cycle, the working software should be ready to be presented. This is the right time for the team to go through everything before presenting it to the stakeholders and other guests present.

The best approach to take is to review the initial plan and confirm that all the requirements have been met. As the product owner, the decision is yours to decide whether certain functions should be removed or not. In case you are not satisfied with a given functionality, it is your right to question. Find out how they can adjust their speed of working so that they can meet the targets of the next round.

Who should be present?

The whole project team plus stakeholders must be present during each sprint review. Each group of a participant must suggest their feedback after seeing the product.
When should it happen?

The review of a sprint should occur at the end of each sprint. This should not last more than an hour or two.

Step 7: What is next? Choose what you want to focus on in your next sprint.

What is it?

For Agile Project Management to work correctly, there must be a template or rough draft available to show what should be done next. This should be defined in the next sprint. Once a sprint is complete, that is the right time to outline activities and tasks for the next round. To help determine what should be done in the next sprint, it is important to revise the recommendations of the previous sprint and figure out how to integrate that to the next one.

Who should be present?

Stakeholders, directors, and team members of the project should be present.

When should this happen?

It is good when a sprint retrospective can take place after a sprint review.

How long should that happen?

Make it short and interesting. One to two hours is enough to review the current sprint and briefly plan the next.

What should happen then?

At this stage, it is important to have working software to ship for users to review and send feedback. This is the right time to plan for some new features. An important feature that makes Agile the best practice to use is the endless shipping, building, and learning.

Agile supports the release of a half-baked product and letting users send feedback about the product. This means that one is able to recognize a missing functionality early before the final release of a product.

Chapter 3: Agile versus Waterfall Model

In this modern era, smart project management has risen to be an important tool for businesses. Smart project management is the reason why businesses are running smoothly without experiencing challenges in their processes. So far, every business, whether small or large, is making use of technology and better project management tools to ensure that it builds the right software and deliver the correct software. Regardless of whether it is team collaboration, the use of these tools has enhanced the software development process. In fact, it has made everything run so well with the least challenges.

Although there are numerous approaches to apply when it comes to managing a project, the most outstanding approach is the Agile Project Management, since it is flexible and practical. Agile practices provide one with multiple abilities to perform several tasks. This is one of the reasons why it is very popular. Below are some of the differences between Agile and the traditional approach.

Traditional

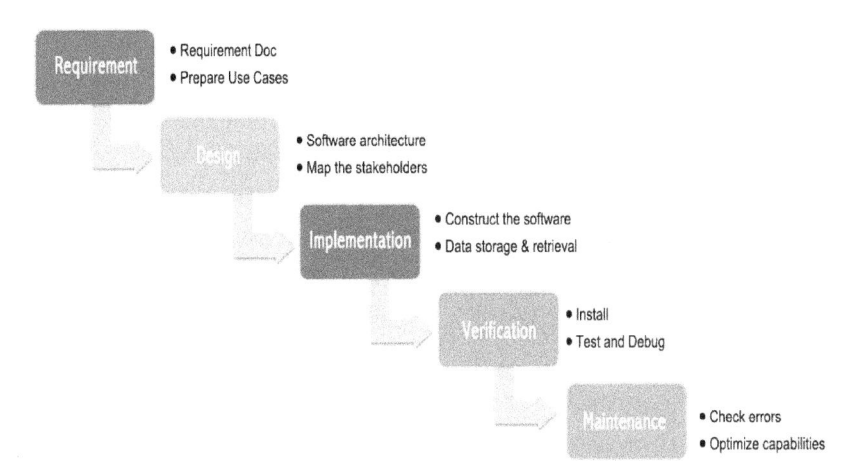

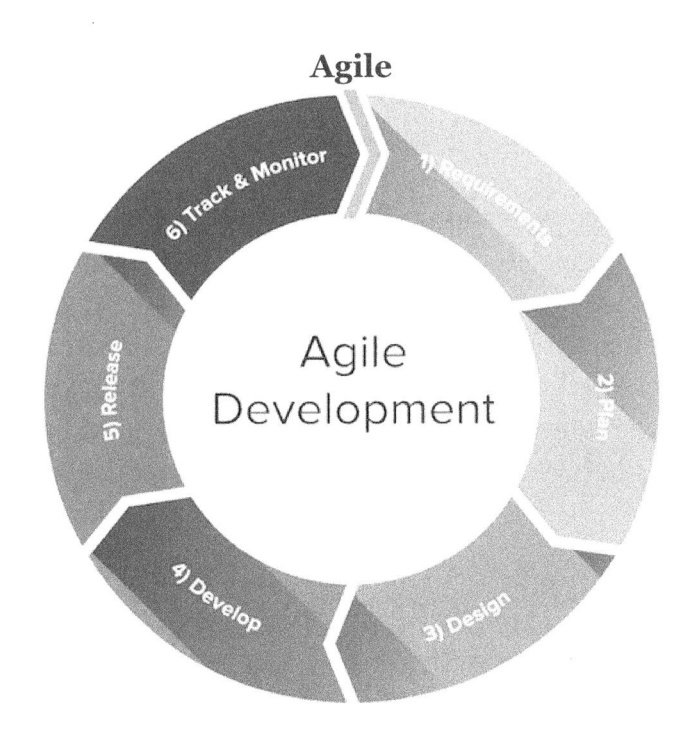

- **Requirement**
  - Requirement Doc
  - Prepare Use Cases
- **Design**
  - Software architecture
  - Map the stakeholders
- **Implementation**
  - Construct the software
  - Data storage & retrieval
- **Verification**
  - Install
  - Test and Debug
- **Maintenance**
  - Check errors
  - Optimize capabilities

Agile

Agile Development

1) Requirements
2) Plan
3) Design
4) Develop
5) Release
6) Track & Monitor

Overview of Agile versus Traditional Project Management

Define Traditional Project Management

A lot has been discussed about Agile Project Management and how good it is for project managers to use in their software creation process. However, to understand correctly how good it is, one must first understand how the traditional project management works.

The traditional approach is also called the Waterfall Model. This name was founded based on its shape. This approach assumes a linear pattern where each stage of the process occurs as a sequence. The basic idea of the Waterfall model is predictability. Predictability refers to the forecasting of experience and tools used. Each project assumes the same life cycle. The stages involved include planning, designing, testing, feasibility, and production support.

With the waterfall approach, planning takes place early without a scope to alter the project requirements. This approach focuses most on the time and cost but the project requirements remain the same. This is one of the reasons why projects created using the waterfall model have budget and timeline challenges. The table below shows a summary of the differences between Agile and traditional project management.

Features	Traditional	Agile
Project Scale	Large-scale	Small and medium scale
User requirements	Defined clearly before any implementation	Interactive form of input
Organizational structure	Linear	Iterative
Client Participation	Low	High
Development process	Life cycle model	An evolutionary delivery process
The cost of restarting	High	Low
Model of development	Fixed	Dynamic
Testing	Done after coding	Each iteration
Requirements	Standard and known at the start	Emerges with rapid changes
Architecture	Generates current and predictable requirements.	Generates current requirements

Why do we prefer Agile and not the Waterfall Approach?

Agile is preferred by a lot of developers and project managers. Let's find out some of those reasons:

- **Project complexity**

Agile

This is the right methodology if an organization wants to find a solution to a complex project. An advanced project might contain different phases that are joined together and each stage might depend on other stages instead of just a single one. This is the reason why most project managers prefer to use Agile for large and advanced projects.

Traditional

For small projects that do not have complex features, the traditional approach could be used best. However, one must still recognize that there may be unexpected changes in the project or certain complexities which can affect the whole process and force one to go back to step one.

- **Adaptability**

Agile

Agile is popular because of the level of adaptability it brings in the project development. Complicated projects have multiple stages that depend on each other. Since Agile methodology provides for adaptability, one can take a risk and change something in a given stage.

Traditional

In the Waterfall approach, the concept is that once a specific phase is done, there is no going back. In short, there is no adaptability in the traditional approach. In case a client requests for sudden changes to be made, it becomes difficult to change. The only option is to navigate back to the first step. This can really waste a lot of time.

- **Scope to receive feedback changes**

Agile

When it comes to getting feedback about a given software product, Agile Methodology is the best in this sector. This is because it has a flexible process that embraces feedback once a product is released to users. Feedback helps improve a product as well as fix any issues. Flexibility in the Agile methodology is the major reason why most organizational managers choose to use it. Software developers who code using this methodology can respond fast to customer requests because they deal with small tasks in the big project.

Traditional

In the Waterfall model, each step is defined in detail before one can begin implementation. This approach cannot deal with sudden changes or feedback that might require quick responses. In most cases, the traditional approach has both a fixed time and budget.

Features of the Agile project development

- **Breaks project into smaller parts**

Agile works by dividing a project into smaller chunks called iterations. Then, the iteration is sent to the customer for review. The success of an Agile project depends on what has been achieved after every iteration.

- **Self-organized**

In Agile development, there is a parallel management model. This is where company employees aren't just managed by a single individual but by a group. Generally, Agile projects have three parts:

  - The owner of the product
  - The scrum master
  - The team

- **Customer participation**

When it comes to Agile development, the customer is the first priority. The customer is very important in the built-up of each of the iteration. The role of the customer is to review the iteration and provide feedback. Once the feedback is out, the right action is taken.

Overall, Agile is the leader in the project management system. When you compare it with other approaches, Agile's features are at the forefront. That is why, it is the leading software project management methodology.

Chapter 4: Learning more about Scrum and Agile Principle

Scrum and Agile are terms common in the field of engineering. They are phrases mentioned by tech engineers. It is like part of their language. However, this language can be frustrating to a person who is not used to it.

Scrum and Agile: What is it?

Things may not be easy when you think of starting with Scrum and Agile. The two words can confuse anybody because one can end up using them interchangeably. However, don't get confused because both words have a different meaning.

So far you know the meaning of Agile. No need to define it here. What about scrum?

In summary, Scrum is a framework used in Agile development. An example to illustrate the difference between Scrum and Agile is the difference between a diet and recipe.

An individual who eats vegetables has a diet prepared based on certain methods and principles. A recipe for cooking vegetables is the framework used to create a diet for a vegetarian. This analogy tries to explain how Agile and Scrum are related.

Who can benefit from using Scrum?

It is wrong for one to think that Scrum was built only for engineers and developers. In fact, this framework can be used to build any other project. One can use Scrum in any type of project whether it is in the market industry or any other type of industry. Scrum is the right framework to use to gather ideas and organize a project team.

Parts of a Scrum

To understand how Scrum works, one must know the parts of the framework and the people involved. The best thing about this is that experience is not needed.

No need to have previous knowledge before you can begin to use Scrum. The only requirement is a place to help create ideas. That could be a whiteboard or software such as Trello. Below are the parts of Scrum.

- Product Owner. This is the person who has the best interest of the user and can say what he or she wants to get in the final product. The product owner is in charge of creating a Backlog. A Backlog represents a list of requirements or tasks for the final product.

- Sprint. This is a timeframe when a team is supposed to complete several tasks.

- Daily Scrum. This refers to the daily updates given by the teams.

- Retrospective.

This should help you realize that Scrum is very easy to learn because you don't need to have special tools. The only trick lies in mastering the lingo and sticking to the guidelines.

Basic Scrum Framework

Remember that no previous experience or special training is required to start using Scrum. You can teach yourself. It is easy to learn the basics of Scrum. The hardest part is mastering the techniques. Experts of Scrum believe that one can learn the rules of Scrum in 10 minutes, but it would take years for one to become an expert in Scrum.

The Principles of Agile

There are 12 major principles that can be used to manage a project. These principles work best with Scrum.

1. Customer satisfaction is the leading goal and it is realized through a continuous and rapid delivery of a product or software.
2. Changing environments or circumstances are embraced at any point in the stage to help deliver the best product to the customer.
3. A service or product is delivered at a higher frequency.
4. Stakeholders and developers work hand-in-hand every day.
5. Both the stakeholder and team members should remain inspired to ensure the best results in the project.
6. Physical meetings are the best to find out the progress of a project.
7. A final functional product is the general measure of success.
8. Sustainable development is realized through an Agile process where both the stakeholders and the development team can maintain a constant pace.
9. Agility is realized through a subsequent focus on

technical excellence and correct design.
10. Simplicity is a key aspect.
11. A self-organizing team mostly develops the right architecture and designs to fulfill the requirements.
12. Regular meetings help enhance efficiency and fine-tune a product.

Best Scrum Master Skills One Should Know to help Steer Digital Innovation in the Company

A scrum master is an important person in the entire Agile process. This person is responsible for guiding the team to apply the best practices and remove impediments. Although there is a "Master", these people aren't masters actually. They are a servant-leader and help serve the product owner and service team. This means that there will be a need for a complete change of the mindset. To help propel digital innovation forward, the best scrum master skills that anyone should have is the ability to change and constantly be ready to improve both yourself and your team members.

Master these 3 Scrum Master Skills and become a Scrum Hero

1. Listen

Usually, Type A personality enjoys taking leadership and direct people on what they need to do. There are times when you may feel that you need to stand up and give directions for things to flow well. For instance, you could be the first to speak and provide directions for the scrum meetings. Maybe you already aware of the best way you can accomplish things, and sometimes you just know.

But the focus of Agile development is defined on the team finding a common goal and delivering the end product. The key thing is to get the team to work both effectively and efficiently in the best way possible. It is hard to realize your goal when you are unable to listen and learn from the team members. It is very critical for one to listen to some of the problems that the team comes up with, the suggestions they put forward, and some of the improvements they would like to introduce.

How to listen effectively

Most people aren't natural listeners and the best way to start is to act like a good listener. Make sure that you don't speak until the other person has made their point. Avoid interrupting a person while they are speaking. Secondly, stop arguing with others silently in your head or plan on how you are going to respond. Finally, try and repeat whatever they said to make sure that you understand everything they wanted to say. Practice your listening skills every day with a one-to-one conversation and then you can advance into the scrum meetings.

2. Coach

The team you guide and direct is the most important organ of the Agile process. Therefore, you have the responsibility to ensure every individual on the team grows and develops. Any time you discover a developer struggling, avoid to brush it under the rug. Come out and help them. Guide them using tips through each problem and help them achieve whatever that they are doing. Make use of a one-to-one time in the meeting. This style of coaching has been found to be important at the start of a project. Also, it will help one avoid problems that might arise later in the project. If you find a person on the team that is causing a given problem, address this issue early. Do not be afraid or shy to talk about it. Remember. A small problem ignored today may turn to be a bigger problem tomorrow.

How to coach people in the most effective way?

The goal of coaching is to generate feedback to team members and individuals. Do this as many times as you can. Has one developer or the whole team done something good? Well, praise them in private or public depending on their personality traits. Do you find something that an individual can improve? Express it to the members. Begin by stating what went wrong, and creative suggestions on how to make it good. When you learn to give suggestions or feedback every time, team members will learn to expect and appreciate it. While feedback becomes a habit, team members will begin to take it personally or even shy away from it.

3. Facilitate

Some of the goals of a scrum master are to facilitate meetings and coach people on the right practices. This means that you are not supposed to direct other people on what they should do. The team should work collectively at every meeting. The best facilitators are those that no one notices that he or she were present in the meeting.

How to facilitate in the most effective way?

One of the major skills of effective facilitators is good listening skills. They ask brilliant questions to help them understand the perspective of everyone and steer the group forward. Good facilitators remain neutral and don't favor anyone's opinion. They have great confidence and believe in themselves. They have this belief that an agreement will be attained and solutions found while in the group. It is important to know that no one was born a facilitator and therefore anyone can learn how to be a great facilitator.

In summary, a Scrum master needs to be a hard working. In addition, it is always a great thing to see a team work together and realize the results in a faster way. Although the above skills are important for Scrum Masters, every individual has the role to facilitate, listen and assist others.

Chapter 5: Turning Your Organization Agile

The big question is how to turn a whole organization Agile. Usually, in an organization that has fully implemented Agile practices, it is the role of the manager to ensure that all members remain committed to their duties and ability. Once every team member is dedicated to their tasks, a greater value is delivered to the customer. In addition, the manager should have complete confidence in the action taken by those in contact with the customer. The manager should also trust his or her team to do things in the right way. Agile is not like a top-down or bottom-up approach. The major focus is on delivering the correct value to the customers. In short, the customer is the boss and not the manager.

On the other hand, the manager in the traditional model has a different responsibility than a manager in the Agile methodology. In this model, the role of the manager is to determine what has to be done and instruct the employees to do it. In addition, the managers oversee that the employee completes the work based on all the instructions outlined. This model assumes that the duty of the employee is to follow instructions and trust the decisions of the manager. The main goal of this model is to generate money for the firm. Therefore, the manager is the boss.

In many organizations, the manager is the boss. Therefore, this makes it hard to implement Agile practices. Any efforts taken are thwarted down because of continuous friction between the managers and the Agile team. For that reason, letting team members adopt Agile practices in these type of organization becomes difficult. At most, it will never be realized.

The reason why partial fixes aren't the best solution

There has been reported tension between the management and the adoption of Agile practices in many organizations. To reduce this tension, it is important to change the role of the Agile team leader so that it reflects that of an Agile methodology. This is done by creating a new job description for the project supervisor. However, this approach only delivers a temporary solution. Below are some reasons why it is not permanent.

In a big organization, there are different managerial ranks. Redefining the job description of the manager means that the friction between the Agile team and the top hierarchy was only reduced by a single layer. It is very hard for the friction to end when the top managers in the hierarchy still embrace the traditional leadership.

Another reason for the looming friction is because top managers in big firms have one purpose, and that is to generate money for the company shareholders and executives. This kind of approach is called maximizing the shareholders' value. An approach that is different from the goal of Agile methodology. Remember, the concept of Agile projects is to deliver value to the customer. Generation of money is the result but not the main goal with Agile practices. For that reason, unless a permanent solution is found, it will always be difficult to implement agility in these organizations.

Why does the top rank dislike Agile?

Is it possible to have the top managers of an organization accept Agile without discussing the goal of the organization? The answer is no. Top managers use command

and control approach to address the creation of large profits and increase the stock price. The right solution is to solve the tension between an Agile team and the top management.

The best way: The creative economy

Top companies such as Apple, Zara, and Google do things in a different way. These companies use the Creative Economy. For that reason, they have redefined the goal of the entire organization and shifted the focus from the shareholder to the customer. In these companies, top executives embrace the customer-value. In short, they are Agile-friendly.

Since they have adopted Agile practices, money is the end result but not the main goal. The major goal is to satisfy the customer. Google and Apple are a great reference to say that Agile methodology has delivered profitable results.

Attempting to resolve the friction between Agile and the traditional management in a rational way is very difficult. It is hard because the traditional role of management has a deep-rooted attitude, views, and values about how the world operates. This contributes to the corporate culture. Past experience shows that changing the corporate culture using methodologies, decisions, and descriptions can be hard.

However, it is important to stop managers from expressing a boss mentality and start to embrace Agility. One must reach out to managers on a deeper emotional level through experiences. This has a chance to cause managers to begin to appreciate the different attitudes, values, and understand how the world operates. The manager should build a positive attitude with the customer.

In most cases, this is not easy to realize because the role of a manager seems to be permanently engraved in the culture of the organization. Part of this culture is to protect a set of processes, goals, and values. This means that even though a manager would wish not to act like a boss and uphold the customer, the existing culture makes it hard to change.

The small aspects of a culture are combined to prevent any efforts to change it. Some individual fixes can try to change it but it will be hard because of other aspects of the organization. It is critical to know that this is not like a house which one can remodel one section and leave it that way. No. An organization is like an interlock of patterns. To shift into Agile calls for five major fixes:

- Instead of aiming to make money for the organization, let the main focus be to satisfy the customer.
- Rather than having those who perform the task report to the bosses, the work should be done in a self-organizing team. In addition, the management should ensure that it does not check to verify if those meant to perform a given task have done so. It should aim to facilitate those who are to perform that task.
- Allow work to be carried out based on the Agile principles instead of the work getting coordinated by the bureaucracy rules, reports, and plans.
- Rather than a preoccupation defined by predictability and efficiency, the major values should be continuous improvement and transparency.
- Elimination of a one-way top-down command and replacing it with a horizontal conversation.

Changing the organizational culture

By choosing to apply these five fixes to help implement Agile in a different organization, the end result is that the corporate culture acquires a new image. Don't forget that this is not easy to do. It requires that the company resources are integrated.

Overall, the best strategy to use that has been found to deliver success is to start with the tools of leadership. Some of these tools include control systems, a vision of the future, and the tools of power.

The Importance of Leadership in Storytelling

The motivational aspect needed to change the corporate culture across the world depends on the storytelling ability. Leadership storytelling is more than just getting things done. It is the correct path for leaders to embrace change. Instead of calling for change by producing propositional arguments that may create a lot of arguments, leaders have an opportunity to build credibility using a narrative. Once they demonstrate a great belief in these narratives, the narratives spread and develop creativity, interaction and transformation.

Storytelling is an amazing tool that can perfectly deal with the challenge of changing the corporate culture. It turns abstract figures into a compelling picture. Remember that every great business has a story to tell.

Chapter: 6 Principles of Agile Plus Agile Manifesto

It is just difficult to think of how software and activity have emerged from "The Agile Manifesto." Before this "Manifesto," software development was not a rapid process. This instance usually led to many projects in the pipeline being canceled because of a change in the business need.

The Agile Manifesto is the major foundation of the Agile Movement. The use of Agile beyond the software development has been achieved in the manufacturing, collaboration, communication, and rapid development of granular features under the control of the general plan.

Agile Manifesto-History

This manifesto was released following frustrations experienced in the '90s. There was a major gap experienced between the delivery of a product and requirement analysis. This led to the cancellation of many projects. It was during this period when business requirements and customer requisites changed, and the result was that the final product was not in line with the present needs. The existing software development methodologies did not realize the demands.

Later in 2000, the Agile Manifesto and the 12 principles were released.

Values in the Agile manifesto

There are four key values of the 'Agile Manifesto and the 12 principles' which support it. Each of the Agile methodologies is based on these principles and values.

1. **Individuals and Participations Over Tools and Processes**

It is the first value in the Agile Manifesto. It places a greater emphasis on the people than the tools and processes. This is because people are the ones who respond to the needs of a business and propel its development. If the tools or process spearhead development, chances are that the team will not actively participate in the changes and this has a chance to not satisfy the customer.

2. **Working software over a detailed documentation**

Traditionally, a lot of time was spent on creating a document to be used in the product development and delivery. Technical requirements, specifications, test plans, documentation, and approvals were required. This extensive list caused delays in the development of software. Agile does not eliminate documentation but it provides a much better way to help the developer know what to do. Agile documents are created as user stories that are enough to help in the software development. The Agile Manifesto has respect for documentation but also credits much more respect to working software.

3. **The partnership of customer over contract negotiation**

Negotiation refers to the point when a customer and product manager sit down to come up with the project details.

In a software model such as the Waterfall model, customers discuss in detail the requirements of a product in the early stages. In this case, the customer acts as a participant in the development.

However, they don't take part in the product creation. The Agile Manifesto provides a platform for a customer to collaborate in the whole process of product development. Therefore, it is easy for the team to meet the needs of the customer. The Agile methods may involve the customer during different periods of the demos, but still, it can involve the end user in the daily development. This ensures that all the requirements are achieved.

4. Response to change over a subsequent plan

The traditional software development methodology considers changes in software as an extra expense. As a result, it does not embrace changes. The goal is to create a comprehensive plan that has a collection of features. In this plan, all the features are marked with the highest priority. Also, there is an extensive level of dependency to help a team work on a puzzle. However, with Agile methodology, because of a short iteration, it is possible to transform priorities from one iteration to another. Additionally, new features are integrated into the next iteration.

And so, Agile provides a positive response to change. A popular example is the Method Tailoring technique. This method involves human beings determining the development approach of a system through different changes. Agile allows a team to change the process and ensure that it mirrors the needs of the user.

The 12 Agile Manifesto Principles

These principles describe a culture where change is accepted and the customer is the center of focus. These 12 principles include:

1. Satisfying the customer through frequent software delivery. This means that customers should feel happy when they see the progress of their projects instead of waiting for the time of release.
2. Withstand changes to requirements in the entire process of development.
3. Constant delivery of software that is working.
4. A partnership between developers and business owners in the whole project.
5. Provision of trust, support, and motivation for all the people involved.
6. Provide face-to-face interactions.
7. Working software is the basis of measurement of the progress attained.
8. Agile processes to promote subsequent speed in the development.
9. Focus on the technical specs and design to improve agility.
10. Deliver simplicity. Create products in the right way to ensure everything runs well.
11. Self-organizing groups should promote better architectures, designs, and requirements.
12. A follow-up meeting to ensure that the right product is created.

The purpose of Agile is to make sure both development and business needs are at par. Agile projects are customer-friendly and motivate the customer to participate in the development. Therefore, Agile has turned out to be the game changer in software development.

Chapter 7: Techniques of Agile Software Development

So far, there is a lot that has been discussed about Agile and the reasons why owners of companies and organizations should embrace Agile development in their organization or company. Also, the internet has a lot of content about Agile. Thus, new players that want to adopt Agile may get confused about what is the right thing to do.

Agile is an excellent methodology to use to build a product or software. It is a flexible approach that empowers individuals who want to achieve success in software development and product development. Below are important features that should be present to make sure that Agile approaches succeed.

- A common agreement on process and goals
- Dedication
- Collaboration among all stakeholders
- Openness
- Willingness to share knowledge

Agile Software Development Techniques

Nonstop Integration

This technique consists of team members working on a product. The members then combine their smaller development with the rest of the team. Each integration is evaluated to determine whether there is a problem with the integration process. If a problem is found, appropriate action is taken to fix the problem.

Test Driven Development

It is a coding process that has multiple repetitions of a short development cycle. The first thing a developer does is to create an automated tested case which measures a new function. Then, a shortcode is produced to pass a defined test, before the new code is refactored to accept new standards.

Pair Programming

In this technique, there are two programmers who work at one station. The first one is the programmer and the other one is the driver whose work is to review each line of code entered.

Design Patterns

In software engineering, a reusable solution is an important aspect of the design process. A design pattern is not complete until it is translated into a code. It is a procedure of how one can solve a given problem that can be used in many different situations. Patterns are a formal practice that a programmer can execute in the application. Object-oriented patterns show relationships and interactions between objects and classes. They do so without describing the object classes and applications used.

Domain-driven design

The idea behind this technique is as follows:

- Align complex designs on a given model.
- Put the project's primary focus on the domain logic
- Start a creative collaboration process between

domain experts and technical team to reduce the conceptual heart problem

Domain-driven techniques are not a methodology. It simply provides a collection of practices which help determine the design and speed up the software projects that deal with complex domains.

Code refactoring

This is a process of modifying the software system without affecting the external features. The most important thing is that the changes improve the internal structure of the software system.

Chapter 8: Challenges of Implementing Agile

Problems are likely to occur, especially for those who will be using it for the first time. This chapter looks at some of the challenges that can be encountered in the process of implementing Agile.

It is critical to understand that it takes time to convince a whole organization to abandon the traditional model and embrace Agile practices. However, once that is done, there are a lot of benefits that come on the way after going Agile. When adopting Agile practices, the right tool to have is a complex one that will help facilitate a successful implementation. Know that Agile requires a few changes to the corporate culture and a system change in the company or organization.

1. **You find Scrum as a great interference to the real work**

    To be efficient, a Scrum Master and all team members should be experienced in handling team projects. It will help deal with issues such as delays and so on. An experience of six months is enough to handle most of the issues that arise. However, an individual with more than six months' experience is at a better chance to deal with all the problems. Great experience provides value and purpose in the development of Agile projects. A person who has worked on several Waterfall projects and gone through a lot of frustration on how projects are managed is the right fit.

    Lacking this kind of experience can prove it hard to deal with the Scrum Master. Remember that CSM training does not have enough weight, and the Scrum Master will not direct the team to make the daily decisions. Scrum and Agile are practical frameworks that have unique details about each project and need to be carefully considered. Hence, experience is an important thing. Most experienced developers complain that Scrum and Agile are ineffective because of having a limited exposure. All in all, a team that has all the relevant people present can be sure to realize success in the implementation of Agile.

2. **Developers who are used to autonomous working might find Scrum unnecessary and it slows them down.**

There is no question that Scrum creates some overhead in the process of development compared to other development processes that have no formal methodology. A scrum is a tool used to control Agile projects. It helps create useful insights in the management of the project status.

Certain projects are better done by a smaller number of developers who work autonomously. For example, the Personal Kanban. This is a good project management tool to use for such projects. But if you want to narrow down into a team made of product owners and developers, it is important to clearly specify the collaboration among the members of the team. This is the time when Scrum is the right solution to use.

Whether you are going to use a collaborative approach or an individual-based technique, everything should depend on the characteristics of the project. Similarly, in a project that depends on the existing solution where subject experts are present, one is advised to use a collaborative approach such as Scrum. Again, one can change the Agile approach if the number of parties used for communication is more than three.

3. **Some efforts in development aren't fit for a time-boxed Sprint.**

This is yet another problem. Some types of development aren't used in normal-size prints. Below is a partial list:

- A new complex user interface design.
- New architecture system.
- Database ETL that needs transformation, cleaning, etc.

Some of these may involve different trials so that it can get something to work. They all face the same problem of conforming to a given sprint-sized effort.

The purpose of a sprint is to facilitate testing and prove that a backlog item can work properly. Additionally, a sprint supports the creation of the correct functionality. Teamwork, discipline, and attention are required so that no one can extend the delivery of a sprint on the set date. One of the major problems is related to the term "to the end user." If end users are defined as consumers of the application, then certain development tasks may take longer. However, there are some things that one can do to ensure that these tasks are completed using the right Agile framework. Below are the three problem areas to focus.

- **New Architecture system**: This has different hardware parts, software applications, a different organization's IT layers, and administrative staff. It is important to buy hardware, install, and ensure that it works. Applying for third-party hardware requires that in-house applications are accessible to the application. Security should be improved depending on the existing infrastructure of the organization.

- **Complex UI Design:** This can take different trials before getting it right. Both the development team and SMEs should do a lot of trials and check for errors. The trials and errors should lead to a release of different mockups, wireframes, and graphics.

- **Database ETL:** This may require a lot of layers to facilitate cleaning, data extraction, and data transformation. Finally, the data is presented based on the project requirements. A presentation allows the user to see the work output.

Chapter 9: Agile Methodology

Nowadays, most software firms use Agile software development methodology in the best possible ways that they can. Whether it is the first time in the era of application development or not, these days, most development methods depend on the Agile methodology.

Well, what is Agile methodology? How can it be used in the software development? This chapter looks at several Agile methodologies.

1. Scrum

It is a popular method used in many organizations. Scrum has a highly iterative approach that concentrates on determining the major features and objectives before a sprint. The purpose of Scrum is to reduce risk and add value.

Scrum applies a storytelling tool that defines the way certain features should perform and get tested. The Scrum team then goes through a sequence of sprints to deliver small provisions. To facilitate the proper working of a scrum team, it is good to address the answers to questions that might arise early.

Scrum is different from the Waterfall Model because of its iterative and collaborative approach. One key difference is that the Waterfall approach requires a comprehensive documentation. This type of documentation makes it difficult for one to change certain features that may be unsuitable in a specific environment. When working with Scrum methodology, there should be a regular collaboration between developers and testers in the form of sprint retrospectives. These meetings ensure proper communication and emphasis on something that may not be clear. Besides that, a Scrum Master will always be present to check on how the project goes.

Therefore, the fast iterations make this methodology the best for teams working on a project where customers and stakeholders expect an early release of the working product. This kind of participation helps the team make any necessary changes that may be pointed out by the stakeholders or product owners.

2. Kanban

This Agile methodology is focused on manufacturing. At its core, Kanban may be considered as an extensive to-do list. Not different to Scrum, the requirements of the Kanban methodology are monitored based on their current stage during the process.

However, Kanban is not time-sensitive, but it is dependent on priority. This means that anytime a developer wants to jump into the next task, he or she can do that very fast. This approach has a few meetings to help with the planning. It is not like Scrum. As for that reason, it is important that team members should be careful. In this type of methodology, if the developers' rate of working is faster than those testing, there will be several bottlenecks that would arise. In this case, any individual on the team can join and help in the different parts.

Kanban has a simple transition for the ideal teams. To ensure that the transition to Kanban is efficient and smooth, developers, business analysts, stakeholders, and testers have to meet regularly and discuss. While shifting to Kanban, one should remember that this type of methodology would provide you with the fastest means of productivity in your code. However, chances are that the code might have some errors.

Kanban is the best for small teams or those teams that don't build features that should be released to the public. Besides that, it is a topnotch methodology used in different types of product or teams whose major goal is to maintain bugs in a system.

4. Extreme Programming (XP)

Kent Beck is considered as the creator of XP. It is a popular and controversial Agile methodology. It focuses on the provision of high-quality software in a short period. It builds on customer participation, rapid feedback, subsequent planning, and testing.

The initial recipe of the XP was based on four simple values. Those values include feedback, simplicity, communication, and courage. In the Extreme Programming, the customer works closely with the team vested with development. The role of the customer is to define and prioritize smaller units of functionality called "User Stories." The development team creates an approximation to plan and deliver the best priority user stories. These user stories are presented as working and tested software. To ensure that it optimizes productivity, the practices deliver a supportive framework that will help offer guidance to the team.

5. Crystal

This is by far the most lightweight and easy-going approach. The crystal consists of a collection of Agile methodologies, some of which include Crystal orange, Crystal yellow, Crystal clear, and many more. This Crystal family embraces the fact that every project may require a different customized set of practices, policies, and processes. These differences help fulfill the goal of the project.

A few metrics of Crystal include communication, teamwork, and simplicity. Like other Agile methodologies, Crystal embraces early and regular delivery of a working product. In addition, it promotes adaptability, user participation, and elimination of bureaucracy.

6. Dynamic Systems Development Methodology

This methodology is based on nine principles that rotate around the business needs, active participation of the user, empowerment of teams, and constant delivery. DSDM champions for fitness in the business as the major focus in the delivery and acceptance of a given system.

In this methodology, requirements are listed early in the project. Processes are refined to improve them. Requirements, often called iterations, are defined and delivered within a short time. All important tasks must be done in DSDM project. In addition, not every requirement is highly prioritized. The DSDM framework is independent and has to be implemented in relation to other iterative methodologies such as XP.

Chapter 10: Keys to Successful Implementation of Agile

It is not easy to implement Agile methodologies such as Kanban, Scrum, and many others. There are a lot of challenges experienced as an organization, project manager, and team of developers. Despite these challenges, the benefits of using Agile methodology surpass the challenges experienced in the implementation. Below are some things to consider to help implement an Agile methodology successfully.

1. **Begin with the correct Project**

    Agile methodologies are perfect for use in any kind of project. However, the most successful implementation depends on the type of project. By beginning with the right project, it creates more opportunities.

    Using Agile methodologies on classic projects does not produce positive results. Usually, one might lose control with members of a team that may want to go back to methods that they already used. On the other hand, presenting a narrow scope or highly dynamic scope can prove effective with Agile methodologies.

2. **Define the role of the Team**

    The role that a team delivers when working on a predictive project is different from the team that is working on Agile projects. In the traditional model, project managers have total control over the projects. However, in the Agile framework, the project manager is like the driver.

In an Agile project, there must be great discipline and organization among the team. This is one of the hardest areas to organizations that employ management and control method. Realizing the importance of such a team is valuable. Creating a cohesive team that shares a similar goal can be the right way to succeed.

3. **Approximation of efforts is important**

The most common problem when implementing Agile methodology is the idea that estimation is not that important. While it is not necessary to have project estimates, it is recommended to develop an estimate to show when tasks are expected to be finished. This helps keep the team on their toes to complete the work on time.

If a given task is not realized within a given sprint, there is a chance that there is a mistake in the estimate and that must be corrected. The task has to be subdivided into manageable sections and the level of commitment revised. A flexible management will ensure that each reflects on the tasks which offer the largest value.

4. **Learn and manage the limitations**

Agile methodologies come with certain limitations that must be considered. These include scope, deadline, cost, and issues related to quality. It is okay to negotiate on the scope but maintain limitations related to the deadline, cost, and quality. These methods dictate that a given task should not go past a specific effort. Additionally, a time-box should be created to help make use of the sprints.

It is important that limitations remain the same and not even slightly changed because they are a critical section of the model. Making any changes may let one lose control.

5. Control tension

There are certain organizations that look at Agile approaches as a rapid way of moving things. However, for these methods to last, tension in the team has to be controlled. If you can create a team that is motivated, self-managed, results-focused, and efficient, one can be sure to be successful with Agile methodology. In addition, every team must develop the right attitude towards productivity to realize a positive change.

6. Stick to the methodology

Agile methodologies come with a few standards, rules, and products. Therefore, it's right to adhere to the method correctly. It is always a good thing not to change anything so that you can provide room for experience. If there is anything odd, be patient and offer another chance.

Scrum methodologies create a sequence of stages and meetings that have to be maintained so that the methods can operate correctly. You can move from less to more by applying these methods. However, one must follow the instructions precisely.

7. Quality

Quality refers to improving the speed of delivery and controlling the estimates. It is vital that products are delivered right on time when you use Agile methodologies. One thing that must be ensured is that the products delivered must work. The products have to fulfill what it was meant to do.

For this reason, no one should try to abandon quality in the process of product development until the end is attained.

8. Remember that power without control is nothing

These methods are very powerful. They have the ability to inspire teams to have better results in a short time. Agile methodologies mainly emphasize measuring, analyzing, and constant improvement.

Metrics refer to ways of taking charge of project management depending on the actual data instead of opinions, intuition, and emergencies. Speed, dedication, flow, and compliance refer to metrics that need to consider in improving teams and processes.

9. Optimize on Visibility

This is a key element when it comes to the success of a methodology. Implementation of these methods never happens publicly in some of the organizations. However, it is recommended that this type of implementation should be made open so that both the public and the whole of the organization can see everything that is being done.

It is important to refrain from using private Kanban methods when you have the project sponsor showing up. Gain courage and present it. Explain and make use of the clear advantages. Usually, there is no better friend than a project sponsor getting involved in the management. In fact, this reinforces the idea of transparency when dealing with Agile projects.

10. Handle expectations

Most teams and organizations that take this approach believes that their problems will be addressed and that the client will never have a different mindset. They have this belief that the products will forever be perfect. It is good to understand that Agile methodologies don't provide a one-stop solution to every problem. How you are going to deal with the various expectations right from the managers to your team will define the success of your methodology.

The method might fail to work correctly in the first round, which may make teams feel uncomfortable with certain features of the method. This is very common and should be worried about. Soon, progress will start to show up. This is significant, and the results are going to be positive.

11. Choose the correct tools

Using support tools to implement Agile methodologies helps create success in the implementation process. Creating a centralized support to share information, monitor progress, and deliver project control is very important.

You should be careful not to fall into the trap of using free tools that aren't connected to your organization. Agile methodologies are never implemented by teams that have unprofessional tools. It should be the responsibility of the organization to look for the right tools that will help in the adaptation of a given methodology.

12. Revise and improve the method

Once a given methodology is adopted correctly, the next thing to think about is adjustments. Create reviews that allow one to see what is working and what is not working in the organization. These reviews help an individual make relevant change. However, make sure you do this after you have done some trial with the standard models.

Agile methodologies are adaptable and flexible. This makes them easy to use in different projects in an organization. With some little experience, one can identify imbalances and respond accordingly.

Become an effective Product Owner

In every single Agile project, a product owner is one of the most important people that can make a project successful or fail. This section looks at some of the popular scenarios that product owners face in Agile environments.

To understand and manage stakeholders

One of the most common problems that product owners go through is dealing with stakeholders and some of their expectations. For the product owner to be effective at representing stakeholders to the developers, the stakeholder must have complete trust and believe in the product owner. They have to trust the product owner and represent their needs to the developers.

Well, how can one gain the trust of the stakeholders? The first important thing is to learn to listen and learn from stakeholders. It is an upper hand if you already know the stakeholders. However, if you don't know them, get time to familiarize with them. Begin by introducing yourself, your expectations and roles. The next thing to do is to ask them questions related to their job and needs. Listen to some of their disappointments and expectations for the new solution. Usually, people are open to speak about their processes and share their level of expertise.

Their certain cases where you will encounter detractors, these are people who do not support the new product and aren't willing to change or work for you.

Understand and control the Storyboard

The storyboard has your requirements and what developers should create. When the stories aren't straight or effectively written out, the chances of the project being successful is rare. Therefore, a product owner must be focused on controlling the product backlog and sprints. A feedback submitted by stakeholders should be reviewed in detail. Accept feedback that has been defined clearly and has business value. The remaining feedback should be rejected or examined. To get clean feedback and effective stories, ensure that the stakeholders know what a great story and feedback must look like. When the project starts, get in touch with the stakeholders and walk them through the process of creating a story and feedback. When demonstrating, present them with a link between their stories and what has been created by the developers. As they continue to familiarize and learn the process, you should start to improve the process even further.

Understand and choose the correct Product owner

No matter the type of project, it should have multiple product owners. This will remain a challenge and has a chance to succeed or fail. Multiple product owners are likely not going to be effective. Why? It becomes difficult to make a decision. Stakeholders will have a big problem with whom they should approach. Scrum Masters and developers won't know whom they should address their questions. What if there are two opposing views from product owners? Who is going to be responsible or become the real owner of the product? Multiple product owners in a project create assumptions and conflicting viewpoints. Besides that, it may create unnecessary confusion when it comes to decision making.

That is why it is important to have one product owner for any project. When it comes to choosing the right product owner, pick one who is ready to take full control of the whole product and commit a majority of their time on it. They need to know what should be developed and have a goal of the final product. They should be ready to speak with stakeholders and Scrum masters. Finally, the right product owner is the one who prioritizes the functionalities and features of a project and communicates the reasons behind all his or her decisions.

Remain calm and Automate

A product owner has the role to manage many people alongside their expectations. This is a lot of work that needs more effort and time. There are times when you may feel extremely tired. During such occasions, it is advised to take a break. You can ask others to even assist you. Another way that you can do to remain effective is to automate communication. The most important thing is to learn to ask for support from a higher management when you don't feel fine to make decisions.

Conclusion

Adopting Agile practices can be the most significant change in the project team's management. A successful incremental and iterative development calls for an adaptive and progressive approach to be taken in the project management. In addition, the whole team has to embrace the change and the subsequent improvement that this change delivers.

For any change, it is important to demonstrate the significance of the change to help avoid resistance immediately. The only way to realize this is by delivering the right business and technical results efficiently and quickly. The best way to attain these results is to bring the change as part of getting the work done. Iterative development of the Agile practices is not hard. However, changing how people work is hard. This book provides you with the right knowledge needed to implement Agile Software Development practices. Hopefully, the information about Agile techniques and methodologies presented in this book will help anyone in an organization or to build a project successfully. Furthermore, this will help your organization not only succeed but also thrive in using Agile best practices.

Kanban

The Complete Step-by-Step Guide to Agile Project Management with Kanban

Introduction

Congratulations on getting a copy of *Kanban: The Complete Step-by-Step Guide to Agile Project Management with Kanban* and thank you for doing so.

This book is your guide to learning the in's and out's of Kanban and how you can implement it successfully into your business. Kanban's simplicity is the major factor in how your teams can work together efficiently and produce effective projects in less time and for less money. You may have heard rumors that Kanban is "dead," but what you need to learn is how the world has evolved and implemented Kanban and other methodologies to improve their workplaces and workflows. Not all industries and organizations are suited to the Kanban system!

Are you a manufacturer looking to create a more lean process? Or are you developing software and want to deliver quality product faster? Or maybe you are a healthcare system struggling with inventory management? All of these scenarios are great situations for trying out Kanban. It is a simple, visual system that helps teams communicate effectively and take on only the work they can do within a desired time frame. When you are ready to roll it out to your team and stakeholders read over Chapter 6 to learn how Kanban improves workflow at your workplace so you can garner their buy in.

You are at the doorstep of learning not just what the core principles and practices of the Kanban system are, but how you can implement it well in your company. Read on to learn tips and techniques for creating digital boards, changing practices effectively, and gaining buy-in. Explore different approaches to a method with a long history of efficiency and productivity. If you are ready to save money, time and resources, then you are ready to implement Kanban.

There are plenty of books on this subject on the market, thanks again for choosing this one! Every effort was made to ensure it is full of as much useful information as possible, please enjoy!

Chapter 1:
The Current Status of Kanban

The word "Kanban" is Japanese and its translation means "signal," "billboard," or "sign." The adaptation of the word "Kanban" describes a lean manufacturing process for delivering materials that need to be replenished by means of a signal, which is typically an empty space, empty bin, or a card. The system is designed for restocking materials when they are actually required; when a material has been completely consumed, thus creating a process founded on true need. Using this system correctly enables a company to minimize shortages, accomplish higher inventory turns, and tightly control their inventory.

Kanban development is rumored to be from processes employed in supermarkets in America in the mid-1900's. These stores would replenish products in smaller quantities based on their sales. In the 1950's, executives from Toyota came to visit the United States and took notice of the streamlined process in the supermarkets. Today, a lot of the Kanban process is attributed to Toyota's adaptation of this "supermarket" method.

In the 1990's, almost every business and industry was implementing Kanban lean methodology for delivering materials to their manufacturing line. For companies like Toyota, those that have a manufacturing system with an operator repeating their actions continuously, this was an ideal situation. For other industries, like machine shops or Aerospace, this was more of a challenge. These organizations like Aerospace and machine shops ended up employing other material delivery systems, like a pick-list or kitting system. Pick-list is a list of materials needed for an order that the operator should gather for production. Kitting is when a manufacturer groups different but connected items into a bin to be assembled into a product.

It is important to remember that handling materials are wasteful, or as the Japanese say, "Muda." This is because handling materials do not add value to the end product. The touching of materials does not advance the final product's value. It is necessary to handle the materials for manufacturing a product, but you need to find the most efficient method to keep it minimal and productive. Throughput and efficacy are important for the people handling the materials and the operators on the manufacturing line that consume the materials.

There are five general situations when a Kanban method may not be the best for delivering materials:

1. The operator, or the person on the manufacturing line using the materials to make the product, has to turn around to reach the materials or even walk to get what they need to complete their job. These additional activities create even more waste in the process.
2. Sometimes two or more materials appear to be the same thing to the "naked eye" but they are different and mixing them up in the manufacturing process can cause a major error. If this is the case, like for similar-looking shims, a countermeasure is needed to prevent a mistake.
3. The rate for product completion is too short. This is called the "*takt*" time. When there is a limited amount of time, a greater percentage of the time is required for part selection. This is true even if you have the products well distinguished and nearby.
4. Material usage is not consistent. When you offer a wide variety of products, but each product has a low volume, your bins will stay full. The unused materials are then taking up space and costing you money.
5. Materials must be traceable. Kanban does not allow you to mix loose materials into another bin for another lot. You need to keep a record of the lot used and document the materials not used. This is the best method for controlling a lot through a Kanban process, but it is more complicated than other, more loose processes.

If the traditional Kanban process is not right for all situations, what other choices are there? As mentioned earlier, Kitting and Pick-list are ideas. Kitting; however, is a "bad" word in the Lean community, so instead, think of it as a "Kanban Set." This methodology involves a selection of materials related to the sequence of building a specific product. An AGV or cart would bring the set to the manufacturing line. Because the set would not be delivered until is it signaled, usually by the return of an empty set or tray, it is considered a "pull" system.

A "pull" system means materials and products are delivered when an order is placed. This type of method is beneficial because it reduces the need for storing inventory. Lower inventory levels and reduced costs are the secondary benefits. The opposite of a "pull" system, is a "push" model. The "push" system is a forecast for inventory needs. The company then manufactures to meet the forecast and then pushes, or sells, it to their customers. The problem with this model is forecasts are typically wrong and results in many leftover products.

A product does not need to have just one "Kanban Set." To minimize material handling and movement, the materials could be chosen in smaller amounts. This process also minimizes the number of materials on the side of the line, which opens up more space on the production line. Materials coming in can have limited handling because the "Kanban Set" is placed in a logical location.

As an alternative to having materials on the line at each workstation, the "pick to light" process can be more efficient for the manufacturer. "Pick to light" is a system designed for fulfilling orders or manufacturing products. Normally the operator scans a barcode that is adhered to a tote or carton. These bins are a temporary holding place for materials. The scan reveals the location for materials so the operator can retrieve the necessary items. When the correct quantity of items is selected the operator presses a button to indicate the activity was completed. The operator continues to refill bins as indicated by this system until all materials are gathered for the manufacturing process.

What people have realized since the 1990's is that Kanban is still relevant, in the right situations. In other scenarios, like the ones mentioned earlier, quality, use of space, and labor inefficiency can result from trying to force a Kanban process where it does not belong. Instead, these situations should consider a "Kanban Set!"

Chapter 2:
How to Utilize the Kanban Process in a Non-Manufacturing Setting

There are three primary parts to Kanban, no matter if you look at examples from a Toyota factory in the 1950s or on a Lean practitioners Kanban app. The three elements are board, list, and card. Essentially, the board contains a list, which creates the workflows from the various cards. Each is defined below:

1. Board- This is what houses the workflow. In other project management processes, this is referred to as a "workspace" or "project."

2. List- Also called a "lane," this has a series of aligned cards, usually related to the same part of the production line, and is the title of a column on a board. In other project management systems, it is called the "task list" or "to-do list."

3. Card- A card is found under a list title on a board. This is a product that needs to be created or a task that needs to be done. These are actionable items. In other systems, these are called "tasks" or "to-do's."

A Kanban board is as versatile as an Excel spreadsheet; its applications are endless. For example, if you are about to launch a new product, you can have two Kanban boards, one for marketing and one for development. Marketing would create a board with lists titled "Internal Promotion," "Press Pitches," and "Marketing Ideas." Development would create lists such as, "To-do," "Doing, and "Done." Each department would then create task cards to move from one column to the other as they bring up and complete each task.

While this example above only offers two ideas in a non-manufacturing setting, there are many more applications to explore. But no matter how you want to implement a board, you need to master the basics of moving from just having a list and a bunch of cards to developing an efficient and orderly workflow.

8 Primary Features of a Kanban Board

The features presented in this section all function primarily the same, no matter how you implement it. Some features only apply to an electronic version of the Kanban system, such as a Kanban app, while others apply to both a physical Kanban board and an electronic app. In addition, the names and titles may depend on the practitioner or app you use, but again, the functionalities are always similar.

1. Boards and Lists are Filled with Moving Cards

Being able to easily move cards around is critical to effectively utilizing Kanban boards. It is the most utilized feature in a Kanban model. In fact, the existing cards you have on the board will move more than the new ones you create. In an app, you can click on the card, hold down the button on your mouse, and move your cursor to a new location. This action allows you to move your card from one list to another or to change the location of the card on the existing list. Because this is a feature you will often utilize if you are using an app, find out what process works with the app you have chosen and become familiar with the layout and functionality as soon as possible. For example, LeanKit offers the ability to change list locations. You can have a higher or lower list, and you will need to know how to drag a card between them. Try it out and find where and how you can move your cards. After all, you cannot break it!

Unlike in a manual or physical board situation, you can look back the path each card has taken on your Kanban board. When you move a card on a physical board, you will either have to take pictures or mentally remember where it was to know its journey throughout the process. In an app, technology keeps track for you. In many apps, when you click on the card, it will "flip over" to reveal its backside. Here it will often show you its activity list. Much more efficient than the manual way!

2. People Are Invited To A Kanban Board, And Assignments Or Subscriptions Of Individual Cards Are Outlined.

As with other project management systems, collaborators, clients, and teammates can be invited to the project. An invitation can be extended through the app for access to the entire board or only for an individual card. Some apps only allow you to invite app members to the board while others will allow you to invite anyone by entering their name and email into the invitation fields. After they are added to the board, they can then act on the cards. If a member is added to a single card, they can only act on that card. Typically they can edit a card, comment on them, move or add them. In addition, they can also observe the stream of activity relevant to the board they are a part of. This allows the members to see the project process even if they are not an active part of the tasks.

To assign or share the responsibility of a task you can add a card to a member or user. This prompts the app to send notifications related to the activities for the card. If your card gets a comment, for example, you will receive a notification. If someone else is assigned that card, they will get the notification as well. When someone wants to follow the progress of a card but are not responsible for the activities, many apps provide the option to "subscribe" to it. This allows the member to monitor the activity and receive notifications but not act on the card. On the other hand, if you want to "unfollow" the activities of a card, you can unsubscribe from it. This is a good practice when you want to keep your inbox free from unnecessary notifications.

3. On the Backs of Cards, Include Notes or Communicate in Related Discussions

In a physical setting, your comments are confined to the size of a post-it. You can only communicate enough until the post-it is full. Then you run out of space, and the dialogue comes to a fork in the road. In a technological setting, space is boundless. This is another distinct advantage of a virtual board over a physical option. Now you can jot down everything necessary related to the card.

Typically, on the backside of the card, there are fields for card descriptions, a place to upload related files, and a discussion forum. Also, similar to tagging someone on Twitter, you can mention a person directly in the comment or description by writing, "@-(their username)." To access the back of the card, click on the card itself to flip it over or find the link that lists additional features for the card and select "back of card" or another similar phrase.

4. Cards Can Have Tasks or Checklists Attached

A card needs a checklist because each task is not always a simple situation. In a virtual board, the cards can contain 1 or more task lists or checklists to make the card more functional. Thinking back to the marketing example introduced at the beginning of this chapter, the marketing department had a list titled, "Press Pitches." Under this list, there is a card labeled, "Outreach." On this card there is a checklist containing the following to-do items:

- Create a preliminary email for the pitch
- Complete follow-up communication with additional details
- Deliver media assets
- Confirm coverage
- Publish coverage

In some apps, the front of the card illustrates the status and progress of the checklist, showing a stage of completion as each task is checked off. This way each member can easily see the progress of the card. Similar to discussions and notes, checklists can have specific member's names included by using the same format: "@-(their username)."

5. Limits to Work in Progress Included

For new practitioners, creating epic task lists can be exciting but overwhelming to all involved. This is why several apps provide the option to limit the number of tasks that can be created in a list and offer WIP's, or work in progress limits. WIP's are pronounced like "whips." This restriction can be applied to one or all columns on a board, so you have a limit to the number of cards allowed on the list.

When you know the workload your team can realistically handle, you can set your limits accordingly. For example, if your marketing team can realistically produce 3 pitches per week, then set the limit for "Press Pitches" to no more than 3 cards.

6. Cards can be Tagged or Labeled

A "label" or "tag" allows you to clarify certain details of a card that cannot be immediately determined by its location on the list. Your administrator or app will determine if this clarification process is called "tagging" or "labeling." Both terms are synonymous at this point in the Kanban process.

For example, if a marketing idea is for an online publication and not for a print campaign, you can add a label to the card to denote that it will be published electronically. Maybe a card requires outside assistance from another department or one task is more challenging than another; both are situations where you need a clear label. Tags are created uniquely for each board you operate. Change the label's names and colors to fit the workflow for the board you are working on.

7. Due Dates can be Assigned to Cards

When there is not a due date on a card, it probably will not get done. Deadlines are essential to getting tasks completed. Depending on the app you are using you can click on the clock icon on a card or find the field for the deadline on the back of the card. Typically there is a drop-down menu that allows you to select the date the card must be finished by.

Besides setting the date, the people assigned or subscribed to the card can also get notifications when the deadline is approaching and when something is considered overdue.

8. View a Calendar with Cards

Another added benefit of an electronic app is that it offers a seamless calendar view related to the board. In an app, the ability to switch from the "standard" board view to a calendar view is a simple toggle of a switch. This view will show all the advancing deadlines, schedules, and delivery dates for tasks. In this view you can also edit, move and add cards. If something is overdue or cannot be completed by its original deadline, you can drag the card to a new date to reset the deadline.

Trying Out Kanban

If you are interested in implementing the Kanban process in your business but want to test it out first, below are some ideas on how to start small before applying it to your whole company or department:

- Use a Kanban board for your personal to-do list
- Customize a calendar for editorial content
- Create a space to house ideas and content for projects
- Share a plan of action with teammates or clients
- Follow a sales funnel
- Develop a tracking system for applicants to streamline your hiring process

Chapter 3:
Applying Kanban to Lean Manufacturing

Lean manufacturing and the Kanban process are often considered a natural pair. When a manufacturer wants to remove or reduce waste in their process, they use a methodical approach, which classifies them as "Lean." Because Kanban is a method for systematically replacing materials when needed, it is obvious why the two work well together.

The Function of Inventory Management in a Kanban Environment

There is a balance that the Kanban system looks to achieve between having the least amount of inventory possible and functioning at full capacity. This simple concept introduced by a supermarkets restocking process led to the introduction of Toyota's 6 principles of an efficient system:

1. The downstream remove materials in the exact amounts outlined by a Kanban system. "Downstream" can refer to customers, line operators, or anyone coming into contact with supplies or materials.

2. The upstream delivers materials in the exact amount and succession outlined by a Kanban system. "Upstream" can refer to the supplier, manager, or materials handler.

3. Movement or production does not happen without a Kanban task.

4. Every moment and every material should be part of a Kanban list or card.

5. The proceeding downstream should never receive incorrect or defective materials from their direct upstream.

6. The quantity of Kanban processes being used prudently lowers the levels of inventory while also improving the identification of problems in the current process.

In addition, the inventory actually being utilized is aligned with the need for that inventory. "Pulling" is often the term applied to this concept. A signal is sent when a certain material is exhausted. This signal tells the supplier it is time to send more products and consequently an order is tracked in the cycle for replenishment. This simple method also tracks the frequency of necessary restocking. Cards or bins are used to signal the need for refilling specific products.

In Lean manufacturing, bins are a popular method for tracking. A bin process provides a visual indication to start the process of restocking. An operator or employee is given two bins to work from. They are to pull materials from the first bin until it is exhausted, and then they move to the second bin. When the employee moves to the second bin, the first empty bin sends a signal to the line manager to reorder materials. In an efficient system, the employee will be replenished with materials before the second bin is depleted. To decide how many materials should be placed in a bin, first determine how long it will take the supplier to deliver materials and then how long it takes for your operator to deplete one bin.

Pros and Cons of Kanban in Lean Manufacturing

For lean manufacturing, using a logical process for inventory monitoring and customer demand fulfillment makes rational business sense. That is why the Kanban system makes so much sense for this type of application. Despite it being a sound pairing, some considerations must be addressed before a concrete process can be implemented. Recognizing the several pros and possible cons for Kanban lean manufacturing allows your business to implement a Kanban process effectively.

Pros of using a Kanban Inventory Management Process in Lean Manufacturing

1. **Lowers the costs and levels of inventory.** Workspace is increased when there is less inventory cluttering the area. Also, providing the minimum quantity of inventory saves money. The business does not purchase materials that will not be used.

2. **Need is determined by the demands of the customer.** When materials and products are exhausted, you can identify best-selling products. If a product hardly ever needs new parts restocked, you can assume it has a low demand by your customer.

3. **Production is to deliver, not to store.** The line only gets the materials necessary to deliver what is needed. The saved storage space now opens more room on the line for assembly. Also, fewer mistakes are made in grabbing incorrect products because they are being stored on the floor and not a storeroom.

4. **Progress reports reach managers organically.** If your company is using Kanban software or apps to monitor the process, many provide analytics to illustrate the volume of products being constructed and the time frame required for completion. A more integrated Kanban software system can provide a variety of reports to help with improving, organizing, and planning the workflow accordingly.

5. **Decreases archaic inventory.** Excess inventory causes a lot of extra work and consideration for a manufacturer. In addition to finding storage space for it, the company must determine how long to hold on to it, and what to do when it comes time to get rid of it. For example, the manufacturer can decide to throw it away or give it away or try to sell it themselves. Also, when there is stagnant inventory, damaged inventory is more likely to continue downstream. Finding a problem during shipping is the worst-case scenario, especially if it has been taking up time and space for several months. Once it hits the floor the options for dealing with it are limited. It is best to decrease inventory that is not being immediately handled.

6. **Overproduction rarely occurs.** Pulling only happens when materials are needed. Necessity is determined by the demands of the customer. This process means that all the products are selling, and no excess is created.

Cons of using Kanban's Inventory Manufacturing Process in Lean Manufacturing

Before jumping head first into a Kanban inventory management system, you need to do a few things that take time and consideration. First, you must observe the number of materials already being used. This will tell you the amount of stock needed for reordering. This observation can take a large amount of time, depending on your products. This means bins and material levels in the bins will fluctuate as you respond to the patterns and needs. This observational period can slow down production. Production can also be delayed if you do not factor delivery time properly for restocking bins. Consider a bin filled with seven materials. The line takes about 14 days to deplete the bin. This means the supplier will need to deliver more stock between 10 to 12 days. Otherwise, your production will lag. It is common to have these types of delays or fluctuations during the initial observation and implementation period.

How to Create a 2-bin Inventory Management System

After you have established a firm grasp on your inventory flow and reordering methods, you will need to implement your 2-bin inventory system. It will take a bit of time to work out the "kinks," but after a while, you will get it to work seamlessly. Below is a mathematical system developed by Oracle that can assist you:

$D * A * (L + SSD) = (C - 1) * S =$ *Amount of Materials per Bin*

D = Avg. demand for a particular product each day
A = Apportionment currently
L = Replenishment time required for inventory
SSD = "Safety Stock Days"
C = Quantity of cards
S = Size of board

"Safety Stock Days" refers to a "buffer" of time added between when a delivery should arrive and when it is needed on the floor. This is used in case an emergency arises or there is a problem with the shipment.

Software for Inventory Management

Over the past few years tracking inventory and materials have become easier than ever thanks to the introduction of things like software, RFID, and barcodes. Software has been developed to provide automated solutions to a Kanban process. Using a combination of the two-bin visual system with an integrated Kanban software system, your production line, and inventory management has the opportunity to become efficient and profitable. Integrating and automating a Kanban process helps direct your entire supply chain. Besides watching for the signal and restocking materials when needed, the system can track lead times and replenishment times. It can also alert you if a material will be replenished in time for the line or if it is going to cause a delay. The system provides reports about how well the line is producing and what products are selling. Even in a lean manufacturing setting the Kanban production process crosses several departments, so it is important to manage it properly and allow stockholders to view the process.

Chapter 4:
Applying a Kanban Process to Software Development

Software development has been using various project management philosophies for decades now, but that raises the question why you would choose a Kanban system over other methodologies, such as SCRUM, or resources, similar to a Gantt chart? To begin, you need to review the differences between the main methodologies.

Kanban vs. Other Methodologies

SCRUM is still a popular project management process used in software development today. While it is another successful, agile project management approach, there is a major difference between the Kanban process and SCRUM.

For example:

- A Kanban system does not include time boxes like SCRUM requires.
- A Kanban system has fewer tasks, and they are larger than SCRUM tasks.
- A Kanban system does not assess the process often, if at all, as it does in a SCRUM environment.
- A Kanban system only considers the average completion time for a project instead of basing project time on the "speed of the team," like in a SCRUM setting.

For practitioners who are used to the SCRUM environment, thinking that a project is made up of the team's speed, increased dimension, and scrum meetings, may find the idea of removing them outrageous. Those activities are the primary methods for controlling development in a SCRUM system! The real problem with this concept is the illusion of control. Managers are constantly striving for this control, but the reality is that they will never obtain it. A manager's supervision and influence only work if the team wants to work. If they collectively decide not to push for a project's completion, it does not matter what the manager does; the object will fail.

Imagine a different scenario: one where people have fun at work and work efficiently. Managers then need not control the environment. Control in this setting would actually disrupt the situation and raise the cost. In a SCRUM setting, controlling measures add costs by requiring constant discussions, meetings, and time commitments during the changing of sprints. Most sprints require one day to wrap up and one day to start the next. Those extra days could be considered "wasted" opportunity. If you look at it as a percentage, a 2-week sprint requires 20% of the time to be spent in preparing and wrapping up. That is a lot of time! In some SCRUM environments, as much as 40% of the time can be spent on supporting the methodology and not on completing the mission.

The Kanban system; however, focuses on tasks. This differs greatly from a SCRUM process. SCRUM practitioners want a successful sprint. Kanban practitioners want completed tasks. Tasks in a Kanban system are approached from start to finish, not bound to a sprint time frame. The completed work is presented, and the project is deployed based on when it is ready. Tasks do not have an estimated time for completion set by the team. The reason is that there is no need for this time estimate, and an estimate is often wrong, anyway. If a manager believes and supports their team why would an estimate be necessary? Would not the team want to produce the best they can in the fastest time possible? The manager instead focuses their own attention on developing a pool of prioritized tasks. The team's focus is on completing as many tasks as possible from the pool. It is that simple. Control measures are unnecessary. Managers add items to the pool and reorganize the priorities of tasks as needed. That is how a Kanban practitioner runs a software project.

Sample Kanban Lists for Software Development

Sample lists or columns on a Kanban board for software development include:

1. *GOALS*. Useful but optional. Major goals can be placed here. This is more for a regular reminder to the team than it is an actionable list.

2. *STORY*. On-deck tasks are placed on this list. The card on the top is the highest priority. The team then takes the top card and moves it to the next column, typically labeled "develop."

3. *ELABORATE AND ACCEPT*. Other descriptions can be used for this column before they proceed to the "done" list. Each team and workflow differs from one another. Anything that is uncertain for the team's approach, like an approach to a code that is unfinished or designs that are not determined, can be assigned to this list. The team then needs to discuss and decide on how to handle it before moving it to the next list.

4. *DEVELOP*. Until the task is complete, the card stays on the "develop" list. It does not matter what needs to be completed for the card or task, but if there is anything that needs to be done before the task is complete, it is placed on this list.

5. *DONE*. Once a card is on this list, it indicates that the task

has been completed. It signals to the team that nothing further is needed for this task.

This suggested layout means that any list or column can contain a high-priority task. Each task on the board should be completed as soon as possible. Sometimes there is a column made for only high-priority items. This could be in the "goals" column, or another column labeled "expedite." This signals the team that all items on the list need to be completed first. This means anything placed on this list are of the highest importance only. All other items should be placed on the "story" list until "expedite" items are finished.

Other interesting features on a Kanban board are the numbers located under each list or column. These numbers determine the number of cards, or tasks, that are permitted to each list. This is also called "work in progress" or "WIP's." These numbers are not scientifically determined but are rather chosen by the manager according to their discretion. Typically the number is related to the number of developers assigned to the list. It represents the capacity of the team for work. For example, if there are eight developers, "develop" may be given a number four. Putting two will result in bored team members, and they will talk to each other too much. Putting eight will result in each person working on their task but staying on the list too long. This means the focus on the Kanban process is forgotten; the length of time spent on a task to be completed from the start until the finish is not shortened.

The Benefits of a Kanban System to the Software Development Team

There are several reasons a Kanban process is beneficial to your software development team. To start, several tasks can be completed at one time, cutting back on the time required for completing each of them. The actions taken are only the necessary ones, so there is no switching frameworks or tracking various articles. Planning is unnecessary. Tasks are developed when the project begins, not before. Next, "showstoppers" or problems in the process are identified and solutions are found together as a team to keep the process moving forward.

For example, if a task requires the assistance from another department, but the other department is working on another series of tasks, the production of the project must halt. But an efficient Kanban team recognizes the need for teamwork and will band together to solve the problems so all departments can continue to function efficiently. Finally, it is possible to calculate the completion time of average tasks. Cards can be tracked according to their initiation date, all movements on the board, and when it was completed. Even wait times between each step can be averaged. This information can be valuable to a manager's calculations for planning purposes.

The Rules of Kanban

There are only 3 primary requirements in a Kanban setting, even for software development:

1. Production is visual:

    a. Tasks or cards are created to divide work. The cards are then added to the board.
    b. The board contains lists presented as individual columns. Cards are located in a column in order of priority.

2. Each part of production works to minimize WIP, or "work in progress," which refers to the maximum amount of work being completed concurrently.

3. Shortening the amount of time spent on the process is the purpose of consistently working to develop the system and determine the average time for a task to be completed. This time is determined by measuring the time for each step in the production cycle.

These are the only three requirements of a Kanban system! SCRUM contains nine primary requirements. XP requires 13. The traditional RUP process has over 120! This alone illustrates the major difference and benefit of Kanban in software development.

Chapter 5:
How Kanban Reduces Risk and Creates Improved Software

It is amazing that a pull system created by a supermarket and adapted to a car manufacturing company can help software development companies create quality products with reduced risk, but it works. The largest difference is that instead of pulling physical materials from bins, the Kanban agile project management system improves organizational throughput in a software environment. Tasks are "pulled" into the work pipeline when it is required. Schedules and forecasts are not what "pushes" it into production.

Throughput is improved in a software system by:

- Reducing WIP, or work in progress
- Every development phase is unassumingly observed
- Predictability of an organization is improved through metrics and reports
- Minimal impact results from steps towards change that are incremental, evolutionary, small, and continuous
- Capacity to self-manage is developed by motivating and increasing opportunity for the teams
- The actual management of tasks and knowledge of work processes is promoted through the team
- The risks and issues facing the team are discussed objectively and rationally

Software teams accomplish these actions by following a Kanban process.

Work is Visual

The board is the tool used to show the various stages of work at a micro-level. Using this tool highlights the problems and roadblocks before they become a devastating "fire." At a very basic level, the board contains three lists or columns. They are "to do," "doing," and "done." By showing them in columns, the team can easily identify what is left in each phase. This process simply shows the progress of a project without having to update a stakeholder manually by other means like a call or email.

Visualizing the work assists the organizational throughput by:

- Breaking down individual steps from "A" to "Z"
- Every step includes a column, or list, to facilitate a smoother pipeline execution and delineation
- Easier at-a-glance monitoring is thanks to the color-coding assigned to various types of work if used
- Work status is available in a central location to inform relevant people of the progress of the project

Cards can be moved from column to column easily, updating team members in real-time so they can swiftly act on any problems or challenges early on.

WIP is Limited

Using up your time multitasking is something to be avoided in an agile environment. Taking on several tasks at one time opens the developer up to making more errors, each deliverable takes more time, and the cycle for delivery takes longer. The limits placed on the WIP means nothing can be added until something is removed and the limit is chosen in relation to the capabilities and capacity of the team. When each phase is able to take on more work, it is pulled from another place, not pushed on top of existing expectations.

Projects are divided into smaller pieces, and each is tackled independently in a Kanban process. This makes workflow steady and swift. Limits clog the pipeline, but for a valid reason. When something clogs the system, the team is in charge of "cleaning" the clog before adding in new work to the pipeline.

WIP limits assist organizational throughputs by:

- Maintaining time management through an effective structure and system. No matter the size of the team or how complex or simple the project is there is an unvarying process for each task to be finished in a limited timeframe.
- A smooth workflow removes waste from your schedule, resources, and cost. In addition, it eliminates work that is unnecessary. This is the primary function of limits for WIP.

Changes are Incremental

Changes that are completed incrementally develop your current actions in an effort to continually move the project forward. Identifying the joints in the system that cause work to back up is another benefit of a Kanban system. WIP limits are engaged in an early stage to complete work fast and inject new work in a controlled manner. As work moves to each new stage, the limits placed on WIP are refined and the phases change, the improvements to the workflow are more obvious. "Evolutionary" is the term applied to the Kanban methodology because the approach to the work is completed in bite-sized increments.

Incremental changes improve throughput by:

- Constant improvement accomplished at each stage of the process. All the way until the final, "done" step the results are quality deliverables and final outputs are less prone to errors.

Flow Enhancement

In a traditional environment, sometimes the developers can lack the understanding of what to do following the completion of the task they were assigned. This confusion probably occurs because the earlier work had a bug found in it. If this occurs, the developers do not understand what their next course of action should be because the work is considered finished and they have already pulled new work up and updated the board. Implementing a Kanban system prevents this from happening because the next work task that is pulled onto the board and placed in a list is the highest priority task in the backlog.

Following this process lessens crisis and confusion that frequently occurs when there is an unavailable resource. "Done" work can only occur when the product is finalized and in use. This process saves time and reduces the quantity of re-work necessary.

Organizations that adopt a Kanban system find that the time and effort put into the adoption phase is worth it. The challenges of trying a new agile system like this reap several benefits, including reducing risk and improving the outcome. Employing the mechanisms outlined above in this chapter illustrates the way Kanban assists software development teams, which sometimes these teams can be scattered across the globe, onto the same page. This is feasible thanks to virtual boards and integrated Kanban software. These systems provide easily viewable WIP and strategy to provide tracking systems for project status and individual tasks. These tools are designed to make a Kanban system easier for your team.

Developmental processes are also benefited through the implementation of a Kanban system. Bottlenecks are identified, and workflow is efficiently tracked through this methodology. This means if you are searching for an improved throughput for your software development, a Kanban process provides a powerful solution to your needs. Now you can offer exceptional software to your clients in the best delivery time possible while simultaneously lowering the risks associated with the process.

Chapter 6: Applying A Kanban Process to Workflow in Your Company

While a Kanban process can be applied to manufacturing and software development, many organizations do not engage in physical products. Consequently, industry leaders have taken notice of the Kanban system and discovered application methods for it in knowledge or creative environments. Jim Benson and David J. Anderson led the charge to adopt the Kanban process improvements in their business. These changes were embraced by others and have been used by companies for decades, enjoying similar efficiency and quality.

For example, a marketing department can re-imagine their creative process as a "production" line with requests for features beginning at one point and improving the results coming off the line at the other side. Like in other industries, showing a visual process of work from start to finish allows a business to improve their workflow. It shows bottlenecks early in the process and encourages all the WIP to be finished in the time anticipated.

Knowledge work, like your company's workflow, can follow the four steps of the Kanban process effectively.

Workflow Can Be Visual

Just like in a manufacturing or software development environment, the work and workflow are visualized in a Kanban system. This process allows team members and stakeholders the ability to see the process of work tasks. Results and communication are then improved thanks to the transparency of the process, including all the lines of work, bottlenecks, and "showstoppers." Some teams prefer to have a whiteboard with post-it's while others prefer a digital system for displaying the board. Whatever method your company adopts, the purpose is to show how work is moved from start to finish, no matter how "good" or "bad" it appears.

Work in Progress is Limited

The demand is what pulls a task to another stage. Demand can be from the customer, or it can be from an opening in the downstream. Whatever the case, it is similar to a "just in time" objective. Given a specific time frame, the limit for what is manageable for your team is set for your pipeline's capacity. To illustrate, consider a design team. They are not limited to producing a set amount of deliverables related to what the marketing team can use in a campaign.

Flow is the Focus

The beginning of the end of the project is free to flow when there is a limit placed on the work in progress, and it is a visual workflow. The formation of backlogs can be prevented by the early identification and resolution of the bottlenecks. These interruptions can be resolved before they cause a major breakdown in production. This is true in a creative workplace, just as it is in manufacturing. For example, if the design team has trouble taking on a certain amount of graphic design before heading to web marketing, then it is easy to identify a need for additional resources, training, or alternative limitations. This early identification is critical to three factors: conflicts are prioritized based on the "showstopper," the value to the customer remains high and they stay close to the project, and the investment continues to have a positive return for the company. If the work lags in the process, the investment is tied up, the customer's value decreases, and conflicts are mismanaged.

On a Kanban board, there is a number assigned to each group, column, or list. This number identifies the workload the team can realistically produce in a week or another given time frame. When a list is at capacity, it is considered "blocked." This means that nothing can be added to the list until something moves and creates space for it. A task that is moved must go to another list, including the "done" list. If a manager recognizes that a list is constantly "blocked," they can address the backlog before things pile up beyond functionality. For example, if work is completed, but it is waiting for QA to open up to review the work, and items continue to sit and wait for QA, a manager can see the jam and work to address the issue to get the project moving again.

Improvement is Constantly Happening

Improvement opportunities are identified thanks to the continuous analysis and monitoring that the Kanban methodology requires. Quality, the pace of production, the flow tracked throughout the process is what is used to measure the effectiveness of the team. Being able to visualize the workflow for the company is an immensely valuable instrument used by any business to improve the procedures of their workplace. For the "overachiever," who often takes on more work than they can handle, benefits from the limitations placed on the work in progress. A daily and weekly restriction of tasks makes sure your team members do not try to multitask, ultimately costing you time and money. Instead, they are free to produce quality work in the shortest time possible.

A Kanban process is a simple, agile approach, but its efficiency and effectiveness are hard to rival. It assists teams in operating at a higher productivity level, minimizes conflict, and provides an even distribution of value to your clients. If your organization values improvement on a continuous basis, a Kanban system is a reduced-risk and reduced-cost option to consider.

How Kanban Can Fit Any Workplace Team and Workflow

Maybe you are running an organization with several project management teams, or maybe you are a small company with only a couple people dedicated to an agile project. It does not matter the size of your company; a Kanban system can fit in with your organizational goals. Ultimately, implementing this agile project management methodology will offer you peace and prosperity in your company (and often your own life!).

Tell a Complex Story Through a Simple Board

The first thing you see when you enter a company implementing a Kanban process is a gorgeous board. Maybe this board is a colorful array of post-its scattered across various columns, or it is a sleek online board with interactive cards and lists. No matter the version you see, it is there, and it is beautiful. The aesthetic appeal alone makes it a great tool to select to manage your organization and strategize workflow. Consider the phrase, "a picture is worth a thousand words." Yet a Kanban board is worth more than that! A scan of the board reveals the movement of tasks in the project. Things like completed tasks, the state of each task, and the progress of each task are evident, along with even more information. There will be no struggle to find more information about a project. It is all at your fingertips and visual plane with the Kanban board.

Simplicity Means No Learning Curve

Using a board is a simple process. It does not require extensive training to learn how to operate one. The stages of your project are divided easily, and the board is "decorated" with your tasks in specific categories. Each task is assigned to a team member. When the team member assigned to it completes a task, they are able to place it into the next, downstream list. This whole process is self-explanatory, meaning you can save time by not training your team on how to use it. Once you decide to adopt a Kanban system, it is possible to be up and running as a Kanban organization in just a few days.

Shifting Priorities can be Accounted for Quickly

When a customer specifies what they want at the beginning previous project management systems made the assumption that these requirements are frozen until delivery. In reality, this is not what happens. Priorities constantly change over and over again. When the scope changes, a manager needs to shift priorities with it. Changes are easily managed on a Kanban board. Even small changes can be made quickly and visually on the board. When you are using an electronic board, then even the team members assigned to the changed task are alerted.

Your Workplace Runs with Your Workflow

As you create a visual Kanban organization, your work will begin to run more smoothly alongside your workflow. You will see everything, address issues early, and keep the process moving. The hard work is alleviated with the simplicity of Kanban boards. It is flexible, visual, flowing, and easy.

Chapter 7:
Implement A Kanban System Effectively

Your organization can understand lean management easily because it is such a simple method for improving business activity. To illustrate the simplicity of a lean methodology, you cannot get a better example than the Kanban process. It is a tool that controls the flow of information and materials. Despite its simplicity, many organizations are still confused by the concept. Maybe it is because it is so simple. Implementing Kanban should be easy, but it is often implemented below its true potential. This then leads to the process being abandoned. This is why implementing it effectively is so critical.

Below are rules, guidelines, and considerations necessary to a successful Kanban system implementation. Before you begin the process of implementation, consider the following:

A Kanban process can be:

- A device to communicate from the operation last conducted to the usage point. Or, from the supplier to the customer.
- P.O.'s provided to your suppliers.
- Orders for work to your areas of manufacture.
- A tool for visual communication.
- A method for reducing paperwork.
-

A Kanban system should not be used for:

- Batch or lot or single item production. Something you only create a few times a year should not be managed with a Kanban process.
- Stock designed or used for safety purposes.
- Inventory held by a supplier. For example, consignment or dropshipping is not appropriate for a Kanban system. This situation is not considered a "win-win" lean situation.

- A tool to plan long-term. Traditional management methodologies are best for situations like introducing new products, changes to a customer's usage, and changes in engineering.

To start, within your company, select one area to implement a Kanban system. Begin implementation with less than eight items in this area. Alert your business regarding the implementation and answer questions they raise about how the methodology works. Once the initial implementation is successful and smooth in the one area of your organization, consider adding more areas or items to the process.

Guidelines for Successful Implementation

1. Prior to implementing a Kanban process:

    - A reduction is arranged. If this is ignored than the typical process, a "batch production," continues because the sizes of orders are still large.
    - Production and requirements are uniform or level. Kanban can work for complex situations, but when it is early in the implementation phase, it is best, to begin with, requirements that are more uniform.

2. Suppliers outside your organization need to be certified. The history of the supplier's outside quality is the reasons for not requiring the inspection of their deliveries. This way "on hold" or "rejected" deliveries do not prevent the workflow.

3. Choose a bright color for any Kanban related container, cart, or tote and paint them all. A vibrant green is a good choice if you are stuck on what to choose. This allows all members of your organization to recognize a Kanban tool, especially during the implementation phase. It is also an

easy way to keep track of all the materials necessary for your new system, ensuring all items stay in their correct place.

4. "Supermarkets" are a good tool to use if you use it well:

- A "supermarket" provides a temporary place for a supplier to house items that are between the customer and the user.
- This is best employed when several customers are internal and rely on a supplier that is external. Another reason to employ a "supermarket" is when several customers are internal, and a supplier is also internal.
- This method provides a barrier between the many customers and the supplier, so the supplier does not receive several signals from all the requests.
- The "supermarket" sends the only signal to the supplier that is at the highest priority

The Rules For Implementing a Kanban Process in Your Organization

1. Your customers, suppliers, stakeholders, and entire organization need to be involved in your implementation. Do not even try to launch a Kanban process without their knowledge. Anyone that adds value to the chain of production needs to be included. After all, these are the people that support and report your company as a revolutionary. They must be a part of the revolution, too.

2. The source is the origination point for quality. Customers should never receive a defective product or poor information. Immediate correction is required. Otherwise, you risk losing your customer's pipeline.

3. Support equipment must be reliable. Choose an area to

implement a Kanban system where there is TPM or Total Productive Maintenance.

4. Lead times and setup should be short. Requirements for delivery should occur evenly every month. This means a Kanban system should be focused on parts and products that are consistent. Reduce setup and efforts to minimize the lead-time for raw materials for items that differ each month according to the requirements of the customer.

5. Programs to reduce setup at the supplier level, whether external or internal, should be developed. If they do not have their own program in place, you should assist them with one. Lead times and the capacity to manufacture should not be influenced by the time required to set up. That is the only time when a Kanban process should be implemented.

6. Customers should receive the supplier's materials directly. Non-certified suppliers, or those that still require inspection upon delivery, require the usage point to do the inspection. If this is not possible, then a certified supplier should replace the option in use.

7. Trial and error are necessary to find how it will work best for your company. This is because nothing is fixed. When there are changes to the level of sales or containers or cards are reduced because activities are continuously improved, you need to be ready to make changes to your system. This is especially important during the implementation stage of the Kanban system.

How to Implement a Kanban Process Effectively

Once you identify the area and actions you want to address with a Kanban process to begin, you will want to follow the steps outlined below:

1. Create a visual of what you want to accomplish. Assemble a series of photographs that show off how it should look along with a label that is clear and definitive. Make it so simple that even someone not working in your organization can understand it.

2. Theorize the consumption of the product on a daily basis. Use your own observations, data, or ask your employees about how much consumption possibly occurs. Accuracy is not essential at this point. Keep this step simple.

3. Determine how or what you will use to send a signal. Consider things like cards, containers, spaces, color, and if you will use a digital or manual system. If you are completely new to using a Kanban system, consider keeping it simple and doing it by hand to try it out. Do not get caught up in the fancy, expensive software system unless you are ready to expand your Kanban process to other areas or more items.

4. If the materials are heavy, do not use bins. Instead, use carts that roll. Choose a bright color, like a vibrant green as suggested in the previous chapter. If your colors are vibrant green, choose something different. This process requires you to make choices based on your common sense. Decide what you think is best and try it out. You can always alter your choice if you find it is not the best fit.

5. The quantity of Kanban cards or bins must be calculated. This is done through a mathematical equation:

Kanban # = Daily Qty X Lead Time in Days X (1 + ss) / Qty Inside the Container

- Kanban # - The number of cards or containers
- Daily Qty – The number of pieces utilized each day on

average
- Lead Time in Days – Estimate how many days a depleted material is replenished. Always estimate more days than you think necessary to be safe.
- (1 + ss) – Stock for safety is "ss." Typically this is either 10% or 15%. In the equation, ss will appear as 0.1 or 0.15.
- Qty Inside the Container – Choose a number that will provide you with between 1 to 5 days of consumption. Sometimes the number in a box from the supplier will make sense to your production line while other times it will not. Use common sense to set an original estimate and adjust as you implement the process.

To provide you with an example situation, imagine the following:

- *Kanban # - Unknown*
- *Daily Qty – 60*
- *Lead Time in Days – 5*
- *(1 + ss) – 10%*
- *Qty Inside the Container – 15 items*

The equation will appear as such:

Kanban # = 60 X 5 X (1 + 0.1) / 15

The answer then reveals:

Kanban # = 22

This is the number of cards or carts you will use in your Kanban.

1. Assign roles to team members. Make sure each role understands what is expected of them. For example, a user must understand and agree to your estimations. They must also place the signal in the designated location and participate in the process.

There is also a role for the person responsible for moving signals, cards, and restarting the process. The people in the warehouse also play a critical role that you assign. They are the ones refilling containers or carts with the appropriate materials.

This means that they need to know what to do when they get the signal. Finally, you need to assign the facilitator role to the Kanban process. You typically fill this role; however, you can hire an outside contractor or professional to fill this role or assign someone from your team. This role is present during the entire process and can assist in training and problem solving with the rest of the team. To successfully implement Kanban to your project, this role must be filled with an active participant for a minimum of 1 month.

Basically, engaging your people and providing the tools of the Kanban system is the best way to implement a Kanban process successfully. It can resolve production problems or illustrate deficiencies in your line. It can save you time and money. But if you do not roll it out well, you can end up making stakeholders and clients unhappy, disgruntled your employees, slow production time, and cost your company money. This is why, despite the simplicity of the Kanban methodology, it is important that you take your time and launch it well. Having employees and stakeholders focused on the goal, while your line is producing quality work steadily at a fast pace, and waste is reduced is all worth it!

Chapter 8: Implement Kanban Digital Boards for Production

So you have decided to implement a Kanban process. Maybe it is because you are not meeting deadlines. Or maybe it is because your company has grown and your original organizational strategy no longer fits. Whatever the reason, implementing Kanban boards to help with your production is a valid solution.

As introduced in an earlier chapter, there are a couple versions of a Kanban process you can introduce: manual or electronic. While there are great advantages to having a manual board, with post-its and dry erase markers, there is something extra special to using a digital board. Consider all the planning necessary to set up a Kanban board. You need to consider your project to alleviate your tasks each day. The concept again is simple, but the application, meaning the setup of your digital board, must be done correctly. Yes, there is a "wrong" way.

A basic Kanban board has three lists, presented as columns, titled "to do, "doing," and "done." The cards are listed in a relevant column, the highest priority placed on the top of the list or assigned a specific color. This is your starting point. Maybe you operate a company that succeeds with this basic method, but maybe you need more. No matter your needs, the first thing you need to do is always the same: plan.

The Planning Phase

Setting up a board and haphazardly creating cards in random columns is not an effective strategy for using your board. The planning phase occurs before anything is added to your board. This phase can be challenging or easy, depending on your industry and company design. For example, manufacturing finds the planning phase to be easy because their process is more static, while knowledge and creative industries have a more complex planning process because the needs and items change frequently. Manufacturing creates lists for each process step and assigns a task to a person or team. For knowledge industries, it is necessary to consider the work to be completed and how they view the tasks required to complete the workflow.

Before setting up your board, you need to plan your workflow and create your digital board. In knowledge-based industries, digital is clearly the best because it can include comments, a visual platform for all interested parties no matter their geographic location, and an easy method for adding or editing tasks.

Below are the steps involved in the planning process:

1. Reality should be used to create a map for your procedures. Model it as closely as possible. Goals that are unrealistic are counterproductive to your purpose. The job considering needs to be assigned to a trained and skilled team member. Especially in the beginning, allow your team to have a more malleable time frame. You can adjust this as you observe the process as it improves.

2. Improvement in your workflow is the goal of the metrics designed for your board. Make sure what you add includes the correct metrics to accomplish this. The completion of the project will see several changes, therefore, allow for changes to occur on the board. Nothing is solidified on your board, at least not yet. This is still a planning phase, so everything is still a "work in progress."

3. Problems that often occur, like concealed work or lag time, now need to be considered. Projects often go over on time or turn a new direction because problems constantly arise. This is an important consideration. When you encounter a

problem like an "on hold" task because there is a lag time from the supplier or a team member is working on another task, jot down the information. On the next project, identify the challenge concisely and clearly to help the team overcome the issue for the future.

Simplicity on Your Board is Important

A digital board serves one simple, primary purpose: make your life easier. Overcomplicating the content on the board is, therefore, an indirect contradiction to its purpose. Since the idea of the board and the Kanban methodology, in general, is simple, people will get on board fast, but if you keep adding content and changing tasks, they may have a hard time keeping up. Placing a lot of information on the board at one time can also overwhelm your team. It can be especially overpowering if it is new to the team. In addition, you do not want to surprise the team with tasks that were not discussed with them or columns that are unnecessary. Before adding anything new to the list, talk it over with the team, and do not have more than four columns on the board at one time.

Over-simplicity on your Board is Dangerous

If over complication is to be avoided, so is over-simplification. Creating a bare board means your team is less likely to show interest in it. They will not need to refer to it every day, and will then most likely forget about it. The board should provide a method for communication and teamwork on a daily basis. This means you need to organize it well and update it often. This will make sure there is always something new for your team to interact with.

How to Set Up Your Digital Board

Follow the steps below to successfully set up your digital Kanban board:

1. Determine your Kanban software system for your digital board.

2. Begin with the basic setup of your board with three lists titled "to do," "doing," and "done."

3. Limit your team's WIP to prevent overwhelming them. Too many tasks occurring simultaneously can be crushing.

4. Empower your team to choose the items they will work on based on their abilities. This is the foundation of a pull system and a benefit of many of the agile methodologies, including a Kanban methodology.

5. Organize the planning and prioritizing steps. Using demand to prioritize and select items is ensured when you set this up properly.

After the initial set up of your digital board, observe the needs of your organization and adjust accordingly. Consider also looking online, for example, Kanban boards for your industry to help get ideas on how to best structure your own. Finding examples and adjusting the settings to fit the needs of your team by empowering them to produce efficiently is the plan after the initial set up.

"Good" versus "Bad" Boards

The setup of your board can be deemed "good" or "bad" for a variety of reasons, but no matter how it is classified, you need to understand the difference between the two. What works for one company may not be the best for another. A team can remain focused on the project vision with a process that is parallel for all the developers in a small environment. "To do" and "done" are shared amongst the team, but the "doing" lane is filled with horizontal lines. This layout allows the small team to show what steps they are individually completing in association with the entire team's project. Structuring the board this way is great because board changes are easily made, avoiding having to mess with the "to do" and "done" sections. All the team members involved will appreciate the simple flow.

But what if you have a large team? Trying to keep all the tasks for a large group on one board will result in an overly complicated system. All the tasks that people are working on will make the overall flow seem confusing, thus negating the reason for using a Kanban method. Thin out your process that is placed on the board and make it easy for them to use the tool. This could mean creating different boards for sub-groups or sub-teams or creating a list just for a particular action or group. Again, you need to find a solution that works for your team environment and process.

Chapter 9:
Development Tips for Your Kanban Digital Boards

Below are the top tips for maximizing your digital Kanban board. Refer to this chapter often as you begin and continue your Kanban implementation to make sure you are "checking off" the items on the "good" list and are getting the most out of your efforts.

The "Good" List

To determine if your list is "good," see if you can check off all the items on the list below. If you are missing an item, revisit your planning process to correct your board for your team. You want your digital board to check off all the items here:

- Chose to use a **digital format or a physical** one, based on the workflow for your company and your team's needs.

- Have the **minimum amount of columns** possible. Anything over 7 is overkill.

- All your **tickets apply to the present workflow but also embody the complete process**.

- All your **tickets are "high level,"** meaning not every little task is accounted for, but rather represents a story.

- **Items in your backlog have direct links to tickets.**

- **Name the tickets with clear and succinct labels**.

- You have provided **clear conditions for "Definition of To-Do" and "Definition of Done"** to which your team refers ensuring they meet the expectations before moving to the next step.

- **Balance your workflow** through team-capacity WIP limits and plans for handling bottlenecks and "showstoppers." Keep lists "well fed" so team members always have something to move on to next.

- **Reject items that do not meet the standards.** Not meeting standards can refer to items of poor quality or overly large products or outputs that do not fulfill the "definition of done."

- **Assign a team member to each task in the "doing" list** otherwise the task is back to "to do."

- **You have a system in place to check the "done" items are really done.**

Structuring Your Board for Your Team

The columns on your board represent how a project reaches completion from the beginning creation phase. Each stage of this process is represented and is considered the "pipeline" of your work. A Kanban system prefers blurring the line between "stage" and "state," offering columns with the "state" of the project task, such as "doing," instead of saying something like, "analysis" or "testing."

In a lot of settings, you will know what is happening during the state of the task depending on who is working on it, so it is unnecessary to break it into more granular "stages." Some teams; however, desire and value a few more "doing" options. These additional columns represent the needs specific to the project or team. Common examples of additional "doing" columns include:

- "Ready" or "Next up to be selected."
- "Ready for Analysis" or "Ready to Define"
- "Develop" or "Implement"
- "Integrate" or other dependencies from the outside
- "Test"
- "Done" or "Complete"

Adding these additional columns between "to do" and "done" may be overkill for small companies or certain structures while more complex and large organizations will appreciate the visual progression for each stage of the process.

Notes on the Additional Columns

"Ready of Analysis" or "Ready to Define"

These distinctions are only useful for actions in a specific workflow close to other actions and directly needs to function before another action is taken. In many environments, this is an unnecessary column. Instead, the role of analysis should have its own board. This is because most analysis occurs prior to a products completion. Large amounts of analysis are piled in this column and create delays that are not relevant to the completion of the project. Instead, keep these actions independent, if possible.

"Ready"

This means you have to create a "definition of ready." This clarifies the conditions that must be met before the task is ready to work on. Then when an item is placed on the list, it is placed according to its priority level. The highest priority items are on the top and the lowest at the bottom. Each time an item is moved to this column its priority level must be considered.

"Doing"

These are columns are all about development, but sometimes, especially if you do not have a "Ready to Define" column, the actions in this column require analysis to be completed. If you opt to remove the "Ready for Analysis," include the one-off analysis requirements in this list.

"Test"

This is another optional column that can be a waste of time, depending on your industry. For example, companies that are regulated by outside agencies that will need to review your production before completion or deployment benefit from having this additional column. Other industries that do their own internal testing should consider removing this column and including the actions in the other activities. Often "testing" is done when a task is ready to move to the next location to make sure the product is "good" before passing it to someone else or to the next stage in the process.

"Done" or "Complete"

These are exactly what the terms mean: the task is finished. Your "Definition of Done" is critical at this point because it makes sure your team members only place items on this list when they meet your standards for being considered "complete." For many industries, "done" is when a product is released or is ready to be released. It does not mean it is waiting on something external to occur or the item is placed on hold. The definition will vary from business to business, but it should never provide a place to hide remaining work.

"Integrate"

If your company finds that it completes their activity on the task but it then has to wait for an external action, consider including the "Integrate" column. This is the location your team can place something when it is waiting for this outside condition. Sometimes this "integrate" can occur in the middle of development or at the end, so where the card moves to after "Integrate" depends on your workflow and product.

"Relevance"

While not a direct column, "relevance" refers to the items that the team is working on or planning to work on in the upcoming and near future or for the upcoming release. If it is part of a larger body of work, make sure the actions related to the part selected are on your board and nothing else, so it remains free from noise and volatility.

Chapter 10:
The Difference Between Kanban and PAR

PAR systems are still the most common method for healthcare companies to manage medications and supplies. Hospitals are the systems that use it the most. PAR requires each item to have a level set for it. When it drops below "par", it needs to be restocked. The concept is simple; however, to determine "par," inventory conducted manually and counting in a cycle is required. The supply chain is burdened by the added manpower and cost these actions require because the activity does not add value to the system.

Unfortunately, another common practice is to guess at the inventory levels of an item, not physically count each item. This saves time; however, it is inaccurate and can cause waste. It can increase costs and inventory levels. Understanding the opportunity for error, it should come as no surprise why leading manufacturing companies do not utilize this method, despite similar goals: always have on hand a consistent amount of inventory.

The Kanban process, on the other hand, reduces non-value adding activity, like physically counting inventory, because of its visual nature. Each bin has a set number. When it is empty, it is restocked with that number. While it is waiting to be restocked, another bin is being pulled from. Because of the clear advantages of the Kanban system over the PAR methodology, many healthcare systems are beginning to change their processes to a Kanban system. Professionals involved in inventory management for hospitals and healthcare environments have reported positive results thanks to the extra time they now have to focus on valuable activities due to less time spent in the storeroom and ridding of the need for a daily inventory count.

Why Kanban Method Should Replace PAR

There are seven reasons that a Kanban process is a preferential methodology for inventory control over the PAR system.

1. The practice of properly managing inventory is promoted through a Kanban system, not through PAR. Eyeballing the bins to determine if an item is below par is not a good practice, but physically counting each item would require intense amounts of labor and is virtually impossible, especially in larger systems. Keeping the storeroom orderly and "clean" can be maintained easily with a Kanban process, while it cannot be with PAR.

2. The discipline required to restock inventory is easy to maintain with a Kanban system. A set number for each item is marked on the bin label, making it very simple for the handler to know exactly how much should be restocked for that item. It is easy and accurate each time. This means Kanban methodology can prevent shortages much better than the PAR system.

3. Inventory is lean. Does it sound attractive to you to have 50% less inventory on hand and still meet your customer's need consistently? It should! Imagine all the cost, time, and space savings you will experience! 50% is the average inventory reduction PAR users had experienced when they switched to a Kanban practice without compromising their inventory targets.

4. Improvement and management are easier to achieve with the Kanban process. The amount of time between depletion and restocking can be tracked. This information can then be used to set the quantity in each bin appropriately. In addition, this information can be adjusted easily if the supplier's shipping times change or demand is different. PAR makes this management and improvement hard because it restocks items every day, in undetermined quantities.

5. Fixed quantities for replenishment are possible with Kanban processes. PAR requires the daily counting, costing your team a lot of time. Instead, their efforts can be focused elsewhere because a Kanban system provides a fixed amount needed to be ordered and visually signaled by the empty bin. It is a much simpler process.

6. Trips to resupply are reduced with a Kanban process. Bins are refilled when it is needed, not on a daily basis, as it is with the PAR system. This means your trips to restock a bin is greatly reduced. Some Kanban practitioners have estimated their trips to restock have been cut down by as much as 50%.

7. Since bins are not refilled every day, counting does not need to happen every day, as it is required with the PAR system. Kanban methods keep the process as simple as possible: when the bin is empty, reorder the fixed number for that item. While you wait for it to be restocked you pull from the second bin for the product. You continue to pull from that second bin even if the first bin stock arrives before it is empty.

When the second bin is empty, you move the first bin forward and pull from it while the second bin is being restocked. The cycle continues on a loop, refining and improving over time. Not spending all that time counting while using the PAR system will save your company hundreds, maybe even thousands, of man-hours each year! Imagine the improved efficiency and cost-savings.

Inventory management for all healthcare and hospital systems should use a Kanban system over a PAR system. In fact, any industry that must regulate inventory should consider utilizing a Kanban system over the "eyeball the stock" approach of PAR. The industry and your organization can expect enormous financial savings thanks to the reduction in inventory, a minimal shortage, and improved productivity.

How to Easily Change from a PAR to a Kanban System

Because of the obvious advantages of a Kanban process over a PAR system, you may be wondering how it is best to change processes and also avoid errors and frustration. As with anything new, it is best to roll it out in a simple and clear manner after you get the buy-in from all the people involved, including your suppliers! Thankfully, there are technological solutions out there that you can consider to help make the transition easier for your company.

Spacesaver

On the shelf, in front of the two bins containing materials, there is an RFID tag. When an inventory manager recognizes the front bin is empty, they scan the RFID tag and move the empty bin behind the second, full bin. When the tag is scanned, it alerts the person or department in charge of ordering what needs to be refilled. The reason RFID tags were chosen is that they tend to be more accurate and uses less time than a barcode that is printed on a Kanban card or on the box. Of course, RFID tags and the reader cost extra money.

PAR Excellence

To minimize the manpower required even further, PAR Excellence looks to remove not just counting but also scanning barcodes or RFID tags. Instead of tags, scales are placed under the bins. A weight is associated with a full and empty bin. When the bin reaches the empty weight, it signals the person or department in charge of ordering the need to restock that particular item. As with the other solution presented above, this system adds the initial cost of installing scales in your stockroom, and each scale needs to be calibrated for the inventory item and set up on a specific network. Then you need to monitor and maintain a lot of associated data

Logi-D

Similar to Spacesaver, this solution offers an RFID tag to reorder the item that is depleted. Instead of scanning the tag and returning the tag to the shelf, like Spacesaver, Logi-D has a board located on a stockroom wall, which collects the tags of all the items that are being restocked. When the tag is removed from the shelf the empty tag space is colored red, signaling quickly to your inventory manager that the tag is on the wall for reorder.

Technology is "cool" and exciting, but it is an added cost to consider. Do not jump on any technology "bandwagon" because it is new and looks fancy. Choose a system if you think you need it because you see the value it can add to your organization. If it helps, choose a solution. If it does not, keep it simple with a card or a bin.

Conclusion

Thanks for making it through to the end of *Kanban: The Complete Step-by-Step Guide to Agile Project Management with Kanban*. Let's hope it was informative and able to provide you with all the tools you need to achieve your goals.

Your next step is to observe and plan your transformation. Stop wondering how you can become more lean, agile, and efficient. You just read all about it! Now is the time for action. Now is the time to prepare your Kanban board and visual system to make your life easier and your team happier. Now is the time to lower costs and increase production using a simple and effective method.

While you are planning, get the buy-in from your team, company, stakeholders, and even your customers. Sell them on the benefits of adopting a Kanban system, and stay close to the process, refining as needed, so it is the most efficient system for your business. Remember, the goal of this is to assist your team members in working alongside one another efficiently while also benefiting your company. Keeping your eye on this goal during each decision you make will help with all the changes and challenges.

A Kanban methodology applies to a variety of situations, despite rumors it is "outdated." As it is with new technology, do not jump onto the glossy "bandwagon." You have read the options and reviews. Determine the unique needs of your organization and create a way to make this basic system work for you. Do not forget, before you roll out your Kanban board; compare them against the checklist in Chapter 9 to make sure they are "good!" It is a great practice to do each and every time, or at least until you have a firm grasp on the process. The more and more you use boards, lists, and cards, the better your team will get at running an effective Kanban project and process. As they continue to feel empowered and successful, imagine the positive atmosphere and engaged work environment you will have! Success will come to you in a variety of forms thanks to you implementing this methodology in your company. Congratulations on taking this step towards a more productive future for your company!

Finally, if you found this book useful in any way, a review on Amazon is always appreciated!

Scrum

The Complete Step-By-Step Guide to Managing Product Development Using Agile Framework

Introduction

Congratulations on getting a copy of *Scrum: The Complete Step-By-Step Guide to Managing Product Development Using Agile Framework* and thank you for doing so.

When it comes to improving your team's ability to generate useful iterations of a product in a reasonable amount of time, while at the same time ensuring they have the tools they need to cut out as much waste as possible, there is no better choice than a Scrum framework.

While it may have developed a reputation over the years for being somewhat obtuse, this is only because its approach is so much different than what the average team expects that it can seem arcane without a little guidance. As such, this book is here to guide you through the ins and outs of the Scrum process to ensure your team gets on the road to improved efficiency as soon as possible.

First, you will learn all about the basics of Scrum including its underlying philosophy and what makes it so effective. Next, you will learn about the main event in the Scrum process, the Sprint, why it matters and how it will help improve efficiency across the board.
From there, you will also learn about the key artifacts in Scrum and the Scrum Master and how they all work together to improve efficiency on all sides. From there, you will learn about the practical side of the process including how to make the transition to Scrum as well as tips for success while doing so. Finally, you will learn about the success stories of companies from all around the world that have made the transition and seen great results because of it.

There are plenty of books on this subject on the market, thanks again for choosing this one! Every effort was made to ensure it is full of as much useful information as possible, please enjoy!

Chapter 1: Basics of Scrum

Scrum is a process framework in which team members can deal with a wide variety of complex and ever-changing problems in a creative fashion while at the same time remaining productive and delivering products that meet or exceed expectations. While Scrum is relatively lightweight and easy to understand at a basic level, it can also be extremely complex and take years to master.

It was created in the early 1990s by Jeff Sutherland and Ken Schwaber for use in software development but has since been used in a wide variety of other industries as well. Scrum's greatest strength is that it makes it very easy to determine the overall efficacy of work techniques and product management while also making it easier to deal with the issues that come along with striving to continuously improvement the working environment, team and product.

The Scrum framework is made up of various Scrum Teams as well as their associated rules, artifacts, events and roles. Each of these components then serves a very specific purpose, with the whole coming together to be essential to the Scrum framework's continued usage and overall success. Meanwhile, the rules of Scrum are what bind the interactions between the main relationships, artifacts, events and roles that make the Scrum framework work as effectively as possible.

Scrum uses

While Scrum was initially used to develop products, for nearly 30 years it has been used in a wide variety of industries to do things like:

- Determine viable markets, products and technologies
- Identify products ripe for refinement or enhancement
- Iterating and producing new versions of products or additions as quickly as possible

- Sustain existing operational environments and create new ones including cloud environments
- Renew and sustain existing products

Due to its rapid iteration process, Scrum has been used extensively when it comes to developing hardware, software, embedded software and the like. It has also been used for almost everything else including autonomous vehicle creation, governments, schools, marketing strategies and organizational operations too numerous to mention.

While it was created nearly 30 years ago, as the interactions between environmental, market and technological complexities have grown, Scrum has proved its utility when it comes to dealing with life's complexities on a near daily basis. It has also proven especially adept at improving processes related to incremental and iterative transfers of knowledge.

At its heart, Scrum is all about small teams of people working together as effectively as possible. These teams are extremely adaptive and flexible and these strengths can be maintained regardless of how many teams are concurrently working side by side. These teams are then able to interoperate and collaborate via a mixture of targeted developmental architectures and sophisticated release environments. When discussing Scrum, the words development and develop are used when referring to any type of complex work that may be taking place.

Basics of Scrum Theory

The basics of Scrum can be found in the empirical process control theory which itself is part of the philosophy of empiricism. The basic idea behind empiricism is that knowledge is gained most effectively via experience and making the best possible decision at the moment with the information that is available. To take advantage of this idea, Scrum uses an incremental an iterative approach as a means of control risk and increasing the predictability of the desired outcome. There are three pillars at play when the empirical process is used, adaptation, inspection and transparency.

Transparency: Transparency is vital as it is important that those who are responsible for the outcome of a given process have a clear understanding of how it is proceeding at every step along the way. Additionally, transparency is also important to ensure that anyone else who needs to see what is going on can follow along as well. The end goal is that any observers will all have the same general understanding of whatever it is they are seeing.

One such area in which this is the case is when it comes to having a common language throughout the process that can be shared by everyone who has a hand in it. For example, those creating the product and those looking at the results will both need to have the same understanding of when the project is actually completed.

Inspection: Scrum users are frequently required to use Scrum artifacts as they progress towards a goal in order to determine potential variances that may be undesirable to that goal. These inspections should not be so frequent that the get in the way of the work that is being completed and are instead most effective when they are performed diligently by those who are skilled at inspecting this point of work.

Adaptation: When an inspector finds that some aspect or aspects of the process are deviating more than is acceptable, or that the resulting product will ultimately be unacceptable then the process must be changed as quickly as can be managed to avoid additional deviation as much as possible. When adaptation is required, there are several specific events that take place as part of the Scrum process and they include the Sprint Retrospective, Daily Scrum, Sprint Review and Sprint Planning.

To ensure the pillars of Scrum all work at maximum efficiency while at the same time building trust among the group as a whole, the entire Scrum team needs to live by the values of respect, openness, focus, courage and commitment. Scrum team members explore and learn to embody these values as they work with various Scrum artifacts, roles and events. Using Scrum successfully ultimately requires team members to become more adapt at living these specific values over time. Likewise, it is important that the team feel the need to personally commit to achieving the goals of the Scrum team as well.

It is important that the Scrum team feels supported enough to have the courage to always do the right thing on a project, regardless of how difficult it might seem at the time. If the Scrum Team and its various types of stakeholders can all ultimately agree to be open about the work that is being done and the challenges being faced then mutual respect will flourish and everyone can focus on the work of each Sprint and the ultimate goals of the team.

Members of the team

The Scrum team is made up of the Scrum Master, Development Team and the Product Owner. Scrum teams tend to be both cross-functional and self-organizing which means the member of the team will be responsible for choosing how to accomplish their work most effectively as opposed to being directed one place or another by one person or, even worse, someone outside the team entirely. A team is considered cross-functional, however, if it consists of multiple people who can accomplish each part of the goal thus making it possible to virtually ensure the team never has to rely on anyone else to get the job done. The Scrum team model was designed to optimize flexibility, productivity and creativity.

The Scrum team works on an iterative basis which means they deliver products incrementally with the goal of ensuring there are as many opportunities to receive feedback as possible. Each new version is as complete as possible so, at the very least, a version of the project that is usable to some degree is always readily available.

Product owner: The Product Owner is the member of the team who is in charge of ensuring the work of the Development Team is used to its maximum efficiency. What this means is going to vary dramatically based on the industry of the project as well as the individual Product Owner. One thing that will never change, however, is that the Product Owner is the only person who is responsible for managing any product backlog. These tasks can be given to the Development Team but the Product Owner will remain accountable for them. These tasks include things like:

- Expressing available items in the Product Backlog in a clear and concise manner
- Placing these items in order to ensure they are aligned in such a way that best achieves missions and goals
- Work to ensure the backlog is clearly visible to everyone so the next task to be worked on is clear

It is important that the Product Owner is always a single person rather than a group of individuals, though they may work towards the goals of a group if they prefer. Regardless, if the priority of a given item is going to change, the Product Owner needs to give the okay. In order for the Product Owner to ultimately succeed at their task, it is important that it is made clear the entire organization respects their decisions. Likewise, it must be clear that no one else has the power to change what the Development Team is working on or what their current requirements are.

Development Team: The Development Team is made up of those who actually do the work when it comes to creating something that can be labeled as "Done". This is naturally the end of the current Sprint as it is required to move onto the Sprint Review portion of the process. Only the members of this team are allowed to create the Increment.

The Development Team should be structured in such a way that its members feel empowered to do things like managing their own work and organize as they see fit. The resulting synergy this creates serves to optimize their overall effectiveness and efficiency. Good development teams have the following characteristics:

- They have the autonomy to take Backlog items and turn them into Increments that are potentially useable as they deem fit
- Development Team members are Development Team members, there is no further designation in the space
- Likewise, there are no official sub-teams within the Development Team, they can congregate at will
- All accountability is shared throughout the entire team

The size of the Development Team should be small enough that it can pivot as needed, while still being large enough to complete a reasonable amount of work within each Sprint. Generally speaking, if your Development Team doesn't contain at least three individuals then you will tend to see less interaction overall and thus smaller gains when it comes to productivity. Likewise, smaller teams are more likely to be constrained by the skills they do or do not possess, eventually getting to a point where they can't actively improve the Increment in question without going outside the team.

On the other hand, anything more than 10 people can make it difficult to coordinate everyone effectively, leading to decreased gains as well. Additionally, they can add so many moving pieces to the puzzle that the empirical process begins to break down. The Scrum Master and the Product Owner are not included in this classification unless they are actively working on the Sprint Backlog as well.

Chapter 2: The Sprint

While the organizational style of Scrum is relatively freeform, it still uses numerous events to help add some regularity to the process and also to minimize the need for extraneous events that are not directly prescribed by Scrum. Any Scrum events are going to be time-boxed by nature which means that each will have a maximum possible duration. The Sprint is the primary event in the Scrum process in that it contains all the other events that may take place during the creation of one iteration of a product. After a sprint has started its duration is set in stone and cannot be changed regardless if it is to lengthen or shorten it.

The events within the Sprint can all be seen as a separate opportunity to either adapt or inspect something. These events are especially designed to enable the level of detailed inspection that allows for true transparency when it comes to critical processes. As such, if any of these events are missed or skipped for any reason the end result is a weakened ability for the Scrum team to inspect the process and adapt it accordingly, leading to an overall reduced level of transparency that will hurt not only the current Sprint but all additional Sprints moving forward that are based on the incomplete data.

The Sprint

The Sprint is a time-box that lasts no more than a month during which the Scrum team will generate a completed product that is potentially releasable but certainly useable in some capacity which is referred to internally as an Increment. Sprints should each have the same duration throughout the development of the product in question. A new Sprint should start as soon as the last one ends. Each Sprint will consist of various segments including Sprint Planning, Development Work, Daily Scrums and the Sprint Retrospective.

During each Sprint, it is vital that there are no changes made to the scope of the product or Increment that would make it impossible to reach the current Sprint Goal. Likewise, quality goals cannot decrease during the Sprint, though the scope can potentially be renegotiated or clarified if the Product Owner and the Development Team decide that earlier projections were incorrect.

Each Sprint can be considered a type of project with a horizon of a maximum of one month. Like projects, Sprints are used to accomplish something specific which means that each Sprint will naturally have a goal when it comes to what is going to be built, what the design is going to be like and a general, flexible, plan that will set the Development Team on the right track. It will also have a clearly defined scope of the work to be done and what the resulting increment will be.

It is important to ensure that the scope of the Sprint doesn't end up growing so long that you are tempted to expand its length. If the horizon for a specific Sprint grows too long, the scope is likely to change too much and the overall complexity and risk may change it into something else entirely. Sprints are useful because they are predictable and they are predictable because it is possible to ensure adaptation and inspection but most importantly progress towards the goal in a reasonable amount of time. Keeping the Sprint length limited also keeps costs easier to track on a monthly basis.

Sprint cancellation: While a Sprint can't be extended, there is a possibility that it could be cancelled before the Sprint time-box is naturally finished. However, the only person who has the authority to do so is the Product Owner, though they may listen to anyone else involved in the matter, including the shareholders as well as the Scrum Master and the Development Team.

Generally speaking, the only real reason to cancel a Sprint once it is up and running is if the Sprint Goal becomes obsolete. This might happen if the company changes its goals while the Sprint is in progress, or if the technology conditions or market change overall. Generally speaking, however, a Sprint should only if it no longer makes any sense given the current circumstances. This should rarely occur due to the short amount of time that a Sprint is active for, however, especially as the timeframe is quite short.

If a Sprint is canceled, then the first thing that should happen is that any product backlog items that have been generated are reviewed. If any of the Increments are useable or releasable then the Product Owner will accept it while the incomplete items are returned to the backlog with a new estimate towards their completion. Any work that is not immediately useable often has a short shelf-life and will need to be re-estimated if it will ever be used properly.

It is important to keep in mind that if a Sprint is canceled then all of its consumed resources are lost because of what is not reusable. What's more, additional resources are going to be required before anything useful is generated as the Scrum team needs to go back to square one in order to get back to work. As such, the cancellation of a Sprint is often seen as quite traumatic to a team and should be avoided if at all possible.

Planning the Sprint: Any work that is going to be generated during the Sprint should first be discussed during the Sprint Planning portion of Sprint. This plan should be created in a collaborative fashion and include the entirety of the Scrum team. Sprint Planning should be kept to less than eight hours each month, with shorter Sprints having shorter planning periods as well. The Scrum Master is in charge of ensuring that all of the events take place on time and that everyone in attendants is on the same page regarding desired results. The Scrum Master should also be the one in charge of keeping everyone else working within the predetermined time-box.

A quality session of Sprint Planning should answer several questions starting with providing a clear estimation of what can be delivered from the Increment that the Sprint will be creating and how the work that is required be completed. The Development Team is in charge of deciding what types of functionality will be integrated into the next Increment for the upcoming Sprint. Meanwhile, it is the job of the

Product Owner to discuss the purpose of the current Sprint Goal and note the items in the backlog that they believe would help to reach it and would, thus, be most effective in helping everyone achieve their goals. Meanwhile, the entirety of the Scrum team should collaborate in the understanding of the work the Sprint is doing.

The real input for this meeting should include the past project history for the Development Team, their capacity, the most recent increment of the product and the available product backlog. The number of items chosen from the backlog for this Sprint is going to be up to the Development Team as they are the only ones that can accurately determine what they are going to accomplish during the Sprint.

During this period is also when the Scrum team as a whole creates the goal for the next Sprint. This goal should be the main objective that will be met during the Sprint based on the items chosen from the backlog and should serve to guide the Development team throughout the process of building the next increment.

When it comes to determining exactly how the work in question is going to be completed, the Development team determines this aspect of the process after the Sprint Goal has been created and the next round of backlog items has been chosen. The Development team has complete control when it comes to determining how best to add the chosen functionality to the next Increment. This work will naturally require varying levels of effort and work from smaller groups within the Development team of various sizes.

While this outline doesn't need to be exactingly precise, it does need to be formalized enough to determine the scope of what can be completed during the next Sprint. Additionally, this meeting should break down exactly what needs to be done for the early days of the Sprint, generally broken into units of a single day. This plan should then be presented to the Product Owner who may be needed to help clarify specifics regarding backlog items and offer up trade-offs that can be made if the Development Team has too little or too much to do for the next Sprint. Other team member or stakeholders may be invited to this portion of the meeting with the Development Team go-ahead as well.

By the end of the planning portion of the Sprint, the Development Team should be able to concisely explain how the work will be completed during the Sprint to reach the target goals in questions. If there is any uncertainty regarding this fact then this is where it will be hashed out as after this the Sprint shifts into high gear.

Sprint Goal: The guiding principal behind the Sprint Goal is that it provides clear guidance to the Development Team when it comes to creating the best increment possible. A good Sprint goal is one that provides the Development team with a fair amount of flexibility when it comes to the functionality that is ultimately created during the Sprint. Any backlog items that are chosen should all work to deliver on a single element of the products function, which is often reflected in the goal for the specific Sprint as well. It can also be any other type of coherence that serves to keep the Development Team working together towards a common goal as opposed to splintering into numerous smaller, more personal, goals.

While the Development Team is in the midst of a Sprint, it should also keep the Sprint Goal and what is required in order to see it completed successfully in mind. If during the Sprint, the work takes an unexpected turn it is then the responsibility of the Development Team to speak to the Product Owner to ensure that the Sprint can proceed successfully.

Daily Scrum: The Daily Scrum is a 15-minute event that should have its own time-box during which the Development Team can discuss what they will be working on between now and the next Daily Scrum. This will make it possible to optimize the team as effectively as possible for the work that is to come, while also providing a clear path for everyone to follow to keep everyone working towards the same vision. The Daily Scrum should be held at the exact same time every day if at all possible to allow it to be the cornerstone of the Development Team's workday.

Daily Scrums are vital when it comes to ensuring open communication between the Development Team, often to the point that they remove the need for other meetings entirely, thus naturally increasing productivity as a result. They also make it possible for the entire team to be aware of any impediments to the Sprint as quickly as possible. Their daily nature also ensures the team has the ability to make decisions quickly while improving their knowledge at the same time. As such, it is a key component when it comes to the Sprint's ability to improve thanks to adaptation and inspection.

The Development Team should use the Daily Scrum as an opportunity to inspect the progress that has been made towards the Sprint Goal already and what is going to be done next as well. The Daily Scrum is exceedingly useful as it gives the entire team a place to discuss issues that might make it difficult for the Sprint to be completed on schedule. With the whole team aware of the problem, solving it becomes far more manageable, and will take much less time than would otherwise be the case. It also gives the team an opportunity to reorganize as needed to ensure peak efficiency is ensured and reconfirmed each and every day.

The structure of this meeting should be as fluid as the Development Team itself, and the most important thing is that the end result is effective for those who are using it. Some Development Teams start off each meeting with a list of questions to be answered about the current and future state of the Increment while others are more discussion based. Again, the format that your Development Team chooses isn't nearly as important as the fact that it works for them.
The job of the Scrum Master, in this instance, is to ensure that the Development Team holds their meeting, while at the same time not trying to direct the Daily Scrum itself. It is also the Scrum Master's job to ensure the Daily Scrum doesn't exceed its designated time box. Finally, as this is an internal meeting for the Development Team, the Scrum Master should also work to ensure that other team members don't disrupt the meeting.

Chapter 3: Looking Back on a Sprint and Planning for the Future

Sprint review: The Sprint Review is held at the conclusion of each Sprint as a means for the entire Scrum Team and any relevant stakeholders to take a look at the new Increment together for the purposes of determining what changes, if any, need to be made to the new Product Backlog. Throughout the Sprint Review process, the team should discuss how the entire Sprint proceeded, and what can be done to further optimize value in the future.

Nevertheless, this is not a status meeting, it is more informal than that with the presentation of the Increment serving as a means of fostering collaboration and eliciting quality feedback. Assuming the Sprint lasted a full month, the meeting should be no longer than four hours. The Scrum Master is the one in charge of ensuring that this event takes place and that everyone involves correctly understands its purpose. They should also be in charge of keeping the meeting to a reasonable length in proportion to the length of the Sprint.

Every Sprint Review needs to include numerous elements, starting with the attendance; the review should include the Scrum Master, the Product Owner, the Development Team and any stakeholders the Product Owner deems fit. The review should then start with the Product Owner discussing what items have been checked off the product backlog with this iteration and what still needs to be completed before everything is said and done.

From there, the Development Team should discuss everything that went well throughout the Sprint as well as the problems that they face and how they overcome them. They will also demonstrate anything new that the iteration can do as a result of the Sprint and answer any questions from the wider team about this specific Increment.

The Product Owner should then discuss the current state of the product backlog as well as the new delivery date for the final, final product based on the current progress. From there, the entire group should collaborate on what to do next so that the Sprint Review provides useful input for the next round of planning. This should also include a general review of the way in which the marketplace or target audience for the product might have changed during the last Sprint and if anything else needs to change as a response.

Finally, the entire team should review the new details surrounding the product as a whole as well as the current Increment and what the next Increment will look like assuming everything goes according to plan. Anything new is then added to the product backlog as needed.

Sprint retrospective: The Sprint Retrospective is the last opportunity for the team to check in with one another and ensure that there is a plan for enacting any changes that need to take place prior to the start of the next Sprint. Assuming your Sprint lasted one month, this meeting should last no longer than three hours. The Scrum Master should be in charge of keeping everyone on task for the meeting as well as ensuring that it stays within a reasonable timeframe. At the same time, however, the Scrum Master needs to participate as a peer in this meeting as well to ensure they feel the right level of accountability for the Scrum process as a whole.

The goal of the Sprint Retrospective is for the team to understand how effective the last Sprint was in regard to the tools used, processes streamlined, relationships formed and the people involved. This should include a look at what went well and what went poorly as well to ensure that when the team creates a plan for improvement with the next Sprint it accurately covers the scope of what needs to be done.

During this time the Scrum Master should do what they can to ensure the Scrum Team improves, while at the same time keeping everyone tied to the Scrum process framework with the end goal of making the process not only effective but also as enjoyable as work can be. For each Retrospective, the team should emerge with new ways of increasing product quality through an improvement to work process and possibly a change in what the definition of finished for the next Sprint will be. Assuming, of course, these changes don't create a conflict between organizational or product standards.

By the end of each Retrospective, the entire team should have a clear idea of the improvements that will be implemented for the upcoming Sprint. Ensuring these improvements are implemented is where the adaptation part of the process comes into play. While various types of improvements should be implemented throughout the Sprint as they are discovered, the Sprint Retrospective represents an opportunity to focus on both adaptation and inspection in a more formal and focused context.

Chapter 4: Artifacts of Scrum

A Scrum artifact is anything that represents work or value to provide transparency and opportunities for inspection and adaptation. Scrum artifacts are all designed specifically to maximize the transparency of relevant information to ensure that everyone has the same understanding of every aspect of the current Sprint or Increment.

Product backlog: The Product Backlog serves as an ordered list of everything that will ultimately go into the product. This document will be the sole source of requirements for any future changes that will be made to the product. The Product Owner will be the one who is ultimately responsible for the backlog of the product, including things like ordering, determining availability and generating content.

Regardless of how much work is put into the product backlog at the start of the project, it is still going to be a living document which means it will never be truly complete. In fact, the earliest portion of its development is important as it makes it clear what the most clearly defined requirements as well as those that are most important when it comes to getting a working product up and running. The product backlog then evolves from there along with the uses it will have and the environments it will be used in.

The product backlog should ultimately include all the fixes, enhancements, requirements, functions and features that the product will eventually have in future releases. Items in the backlog should all have clearly defined attributes including things like value, estimate and order. These items also frequently include a type of test description designed to determine when it can be considered finished and fully integrated into the next Increment.

As a program starts to get regular use and gains value with users and the marketplace as a whole it generates additional feedback and, as a result, the Product Backlog becomes much more well-defined and extensive. These requirements never stop changing, which is part of the reason it is important for the backlog to be considered a living document.

It doesn't matter how many different Scrum teams are working together on a product, they should all be working from the same product backlog. Refining the product backlog is the process of adding further detail and cost / benefit analysis estimates to existing product backlog items. This process requires the additional revision and review of the items that are already on the list and the Development Team is primarily in charge of determining when this process takes place. A good rule of thumb is that this should take up no more than 10 percent of the Development Team's time. The Product Owner can also update the backlog and any of its items at any time at their discretion.

As a general rule, the higher that an item is on the product backlog, the more well defined it is. This, in turn, leads to more accurate estimates thanks to the increased clarity provided by all the additional details. Backlog items that are going to be dealt with next should be so refined that they can be easily fit into the time-box of the next Sprint because they are so well-defined which means they will be considered ready for selection when it comes time to plan the next Sprint.

In these instances, the Development team is responsible for all of the estimates, thought the Product Owner is not without a degree of sway as well. The Product Owner, in this case, will help by making it clear what trade-offs are available and answering any questions the Development Team might have.

Tracking progress: Throughout each Sprint, the total amount of work remaining should be able to be easily summed up. The Product Owner is responsible for tracking the amount of progress that has been made towards the end goal and updating this progress report after each Sprint Review. The Product Owner then compares the amount of work remaining at previous Sprint Reviews with the current review to ensure that everything is still proceeding apace.

The resulting determinations should then be shared with the entire team.

There are many different practices when it comes to forecasting progress, including things like cumulative flows, burn-ups and burn-downs. None of this can ultimately replace the raw importance of pure empiricism, however, especially in complicated environments when the outcome is far from guaranteed.

Sprint backlog: The Sprint backlog is the product backlog for a specific Sprint as well as a place for delivering the requirements within the scope of the next Sprint. It can also be thought of as a forecast for the Development Team to all them to start considering what will need to be done in order to deliver the next Increment as effectively as possible.

The Sprint backlog is effective because it can make visible all of the work the Development Team considers necessary when it comes to meeting the current Sprint goal. To keep the continuous improvement rolling in, it should also include at least one major improvement as defined by the most recent Sprint Retrospective meeting. The Sprint backlog should contain enough details to make any changes that are made to it easily describable during the Daily Scrum.

When new work is required, the development team can add to the Sprint backlog directly. When the work is completed, the estimate for any remaining work should be updated as well. If specific elements of the plan are ultimately cut for one reason or another, they can then be easily removed. The Development Team is solely responsible for changing the Sprint backlog during the Sprint. As such, it serves as a real-time, extremely accurate, picture of the work the Team is currently working on and will finish by the time the Sprint is completed.

Tracking the progress of the Sprint: The total work that is left to do in the Sprint should be tracked in real time and summed up in a form that is easily accessible to anyone in the team. The Development Team should be in charge of tracking this work total, and updating it with each Daily Scrum to ensure the entire Scrum team knows the odds that the current goal is going to be achieved within the current Sprint.

Increment: The Increment can be thought of as the Product of all the added value the backlog items that have been completed during the Sprint have generated added to the value of all of the previous Sprints thus far. An increment can be thought of as a body of inspectable, completed work that supports the Sprint Goal as well as the end product goal that the Scrum Team is working towards.

Chapter 5: Scrum Master as Servant Leader

Responsibilities of a Servant-Leader

A ScrumMaster is a servant-leader in that it is their goal to facilitate the needs of all of the members of the team as well as anyone the team serves, which is typically the customer. They should strive to achieve results that line up with the goals of the Scrum Team as well as the larger organization's business objectives, principles and values.

ScrumMaster responsibilities may include:

- Setting up a Scrum framework in the service of the team, not as a way to command or micro-manage.
- Giving the Development Team the tools they need to manage themselves successfully.
- Mitigating conflict by ensuring that any disagreement is seen as a healthy exchange of ideas.
- Ensuring that every member of the team is fully versed in the ins and outs of the Scrum framework.
- Stepping in to handle anything that will get in the way of the Scrum Team reaching the goals of their current Sprint.
- Doing anything that is required to remove any roadblocks that may come up during a Sprint.
- Ensuring everyone on all sides of the team are being as transparent as possible.
- Helping in any way possible that will lead to the team becoming better versions of themselves.
- Nurturing a collaborative, supportive and empathic culture within the team.
- Constantly keeping the team challenged and away from mediocrity.
- Ensuring development, growth, and happiness of team members.

The Scrum Master is a servant-leader for the Scrum team. To encourage Servant Leadership behavior, the Scrum Master role by design does not have organizational authority or power. The Scrum Master is not a boss or an alternate title for a manager of the team.

The absence of organizational power, allows the Scrum Master to establish Psychological safety within the team. This, in turn, empowers the team members and allows them to self-organize. If the Scrum Master possesses organizational power, that limits the chances of establishing a safe environment.

A Servant-leader Scrum Master creates an environment where people can contribute and flourish. An environment where people are cared for and feel safe to express themselves. An environment where they've enough empowerment to make necessary decisions. Scrum Master is that leader for the Scrum team.

Who does the Scrum Master serve? The Scrum Master serves the Development Team, the Product Owner, and the Organization in their endeavor to apply Scrum and get benefits from it.

A Scrum Master enables the Scrum Team to become a high performing team. So much so, that it can rapidly adapt to the changing customer needs and solve customer challenges. For those Scrum Masters who also happens to have organizational authority - aka responsibility of product delivery, the team members reporting into you, you make financial decisions, you write performance reviews, etc. observe your behavior closely.

A good servant-leader Scrum Master should be able to answer the following questions:

If asked, will my colleagues and team members say that I serve them?

Whose agenda do I serve? Theirs or mine?

Am I able to justify the responsibility as a Servant-Leader Scrum Master?

How does Scrum Master's Servant-Leadership style work with traditional managers? The paradoxical style of Servant Leadership is difficult to enact for the traditional managers. Most managers tend to be comfortable with the leadership aspect, but not the servant aspect.

Servanthood and Caring: As a Scrum Master, you do not have any authority in the organization. You derive influence from your subject matter expertise of Scrum and by having the heart to serve your team and care for them.

As a Servant-Leader, you seek to empower the team members and invite them in decision making. Your behavior is of serving and caring. It enhances the growth of team members while improving the caring and quality of organizational life.

Is your emphasis on serving your team-members for their good and not just the good of the organization? If yes, then you sure are one effective Scrum Master.

Empowering and Helping: The Scrum Master needs to be concerned about what is going on with all of their stakeholders. Broadly stakeholders include society, communities, business partners and employees, and specifically the least privileged among them. Servant-Leader Scrum Masters believe that team members have an intrinsic value beyond their apparent responsibilities as employees. These Scrum Masters are deeply committed to the development and growth of each and every Scrum Team member. During their time spent as a Scrum Master, individuals need to learn to nurture the professional as well as the personal growth of team members.

Serving Team's Agenda: A Scrum Master as a servant-leader uses his capabilities and skills to help the team establish their agenda. The Scrum Master serves the team's agenda, not their own.

The Scrum Master does not impose any directions or mandate upon the team. A Servant-leader Scrum Master instead, believes in Change by Invitation. They invite the team to choose the goals and the direction. They invite team members to opt into, participate and keep options open for anyone to opt out.

It is important to understand that if a person is titled as a Scrum Master but also carries the traditional manager's responsibility to deliver a release or to manage the team members etc. They'll not be able to truly serve the team's agenda. Such a person will almost always end up making others follow their direction and agenda that they set.

There is also a boundary to serving team's agenda. For example: if a product owner's agenda is to finish a certain number of features by this sprint, however, the team clearly sees that as not practical. Though you want the Product Owner to succeed, however, in such situation it is your responsibility to shield the Development Team from the excessive pressure of the Product Owner. Often Scrum Masters give-in to the pressure and allow the PO to overload the Dev Team.

What do you think is the result in situations when the Development Team was asked to deliver more than they could practically deliver?

a) A product having defects and poor quality

b) Stressed and overworked Team members from having to work extra nights and weekends

c) Accrual of Technical Debt

In any of the situation, can you say the Team's Agenda was served?

Building Relationships: Establishing and nurturing long-term relationships with all stakeholders and keeping the team-members in focus helps them meet their fullest potential. If you are genuinely serving, caring and helping your team members grow, building relations with them will not be an issue. To build longer term relations, you would need to forgo short term approach/gains and allow for the things to settle.

Healthy relations with the team create a synergy among the team-members and boosts the team's performance and growth. Is your emphasis on building long term and healthy relationships? If yes, you are on track.

Being Humble: Like a good leader, the Scrum Master stays humble and practices regular self-reflection. Counter to a traditional leader's pride, servant leaders exhibit humility in their behavior. Servant leaders don't think less of themselves they just think of themselves less.

They have high self-confidence but very low situational confidence. If they are faced with a situation, their response would most likely be: I have the intellect to solve all the problems, but I don't have all the answers and for that, I need other people's brain.

In today's world where there are so much information and so many tools, it's important to acknowledge that one person cannot know everything and that everyone needs or at some time will need his/her team members' help. A servant-leader will not take pride in the moments of success but will surely accept errors in times of failure.

Emotional Healing: Your people are going through change all the time. There is uncertainty and failures. Some of your people may have bruises. Many of them may go through emotional turbulences. Are you able to emotionally heal them? Offer your support?

As per the team development model, the team goes through Forming, Norming, Storming and Performing phases. While your team is going through Forming, Norming and Storming phase of the team development, as a Scrum Master, are standing by your team during this time of change? As a Servant-Leader, any emotional healing and support that you offer can go long way in building an environment of trust and care within the team.

Being Empathic: Being Empathic involves deeply connecting with the emotions of the other individual without judgement and critique. It is an essential behavior of Servant-Leaders. Empathy starts with listening. Genuinely being present at the moment with somebody and listening with your whole self helps to understand the other person's situation. Here the aim is to slow down and listen with the intent to understand the meaning behind the words, the meaning of what is being felt, and what is not being said. Empathy connects two people by heart. Connecting with someone by heart is much more powerful than connecting only through the brain.

For Scrum Masters who are not naturally empathic (count me in with you), being aware of and caring about others' emotions is the starting point of developing empathy. Empathically listening to what your team members say and acknowledging what you sense + hear.

Such as:

Elena, you seem to be worried. How can I help?

David, I hear you are concerned about Matt's behavior. What would you like to happen?

When you lend someone your empathic ears, they get it. They feel safe and comfortable to share even more. Scrum Master through Empathy builds relationships, heals the team members, earns trust and gains influence.

Being Ethical: The moral component of the Scrum Masters must be strong. Being ethical relates to the way in which a servant-leader makes choices, disciplines himself or herself and chooses the right thing to do in the service of the team. The Scrum Master may also encourage the team to self-reflect and establish high standards of moral and ethical behavior.

The team members constantly observe the moral basis of the servant-leader's actions and organizational goals and relate to them.

If you as a Scrum Master have ingrained integrity and professionalism in yourself, it'll be possible to bring it to the team.

Often times, the Scrum Master may realize that the team needs to mature and they must be empowered, educated to handle their own meetings, hold each other accountable, collaborate with users and PO and deliver value. And if the time comes when the Scrum Master may not be providing the best value for the team and hence should decide to either step down from that role or move on.

Chapter 6: Making the Scrum Transition

While there are any number of reasons, that have already been discussed, that could be the turning point when it comes to your decision to transition your team to the Scrum framework, it is important to keep in mind that it's not without its share of difficulties as well. This chapter will look at many of the most common difficulties often associated with the transition and discuss the easiest ways to avoid them.

Team members resisting change: When it comes to the challenges that are often going to be faced when making a Scrum transition, perhaps the most frustrating for a future Scrum master is the covert, overt, passive and active resistance that you will likely face from the team. While the best way to deal with this resistance is likely going to vary from team to team, it is important to know what to look for so you can nip it in the bud as quickly as possible.

For starters, active resistance is the easiest to spot as it is often limited to a handful of jaded and grumpy individuals who don't like anything that gets in the way of how they do things. However, if this issue is not dealt with on a one on one, personal, level then these agitators could generate a galvanizing message that could spread to other team members until there is an active block to prevent the change from moving forward.

This will generally go hand in hand with overt resistance in which those who are against the change are actively bad mouthing the process and trying to talk even more team members out of participating. About the only good thing when it comes to overt resistance is the fact that it is easy to determine who is doing what so that you can get to the root of the problem as well. When you do deal with these individuals it is important to do it in such a way that you turn foes into friends as opposed to doing something that could further damage morale.

Unfortunately, overt resistance is not nearly as common as passive resistance which is far more difficult to pin down, and far more harmful to the cause as a result. While they may learn what to do, they will only ever do the bare amount of testing that they really need to do in order to get by without embracing the process to the extent that the process as a whole is likely to actually improve. This then leads to an even greater issue as it can mean errors aren't caught and other team members have to waste time covering for the person who resists.

Perhaps the most challenging part of dealing with resistance, however, is that it might be a subconscious response that occurs in a team member who appears to have completely bought into the process at first glance. Remember, just because a team member might know that Scrum is the right thing to do doesn't mean that it is the easiest thing to do, which is where the mental disconnect might come into play.

In order to break through this mental disconnect, the goal should be to create the sort of feeling within the organization that this type of change is inevitable. If team members understand that Scrum is already a done deal then they will be more likely to spend time fighting it and more time learning how to make the change as easy to handle as possible.

Motivation is another key factor in this instance as it is important to create a sense of urgency in the transition as well. Selling Scrum as a way of helping the company which is struggling, whether this is actually the case or not, can be an effective means of quelling doubt very quickly.

Misunderstanding of the process: As the Scrum process is so very different from many of its alternatives, it is extremely easy for team members to become confused, despite their best intentions. In fact, it is quite common for team members to assume they understand Scrum only to find out that they are actually mixing up a host of different, similar, processes.

In order to help get them in the right mindset, the first thing the team needs to understand is that what is occurring amounts to a true culture change for the company that will alter how every member of the team spends at least a portion of their day. One of the most common misunderstandings that are sure to arise is the idea of deadlines as opposed to estimates, which are very different things and can take some getting used to. It is important to keep in mind that true reeducation involves learning to think of production as a process which means learning to think in increments and manage expectations differently as a result.

Meetings: Depending on how the workplace was previously broken down, getting used to the idea of a cross-functional team where everyone understands the project can be quite a lot to swallow. If a strict policy on meetings isn't implemented early then you will find that people still attend far too many meetings each day and the team's productivity suffers as a result.

In order to properly address these issues, it is important that the team does a full transition to Scrum all at once as trying to go slowly will only further complicate the issue. At the same time, however, it is important to understand that the team will likely need some time to get comfortable with the lack of control from the top down and you will need to provide support during this time as well.

Intimidation factor: When teams start managing their own work, it means that each member will naturally have much more responsibility when it comes to decision making, prioritizing and scheduling which can feel like a lot of extra pressure. To counteract this fear, you need to make it clear that nearly all of these decisions are going to be made by the team as a whole. Another way to make this part of the process seem more manageable is to start with smaller teams right off the bat. This will keep the individual number of moving pieces quite low which, will, in turn, make the entire process seem less intimidating right from the start. If you have a larger team and you are starting off with smaller teams to make the process more manageable then it is important to also ensure that they always remain as cross-functional as possible.

Chapter 7: Tips for Success

Make a Point of Not Planning Up Front: Many teams, especially when they are first getting started with Scrum, still feel the need to do some type of planning before they get started. This can, in turn, lead to what is known as analysis paralysis. This occurs when the planning stage of any given project will often grind progress on that goal to a halt as buy-in is obtained from several different sources over the course of days, or even weeks. Don't forget, Scrum was designed to be an adaptable framework for inspection, which means that by its very nature it is antithetical to have several team members sitting around waiting for a single individual to finish preparing so they can get to work.

It is important to remember that in order for Scrum to be used correctly, users need to value the act of creation and, as such, the potential for failure inherent therein, not just the documentation of the process. It also values the individual bonds that ensure individuals work well together as opposed to blindly following a set group of rules. This means that while there is no harm in a certain individual from doing some early legwork on a project before bringing in the whole team, the point that this starts to be a detriment is where waiting for this one person starts keeping a larger number of individuals from generating any real type of value for a prolonged period of time.

While finding the right mixture of pre-production and production can be difficult at first as you don't have a clear idea of what exactly it is you are looking for from your team, it will get easier to pick out with practice. A good rule of thumb is that if it takes you more than two days to fully prepare for the start of a new project, then your team might be suffering from analysis paralysis.

Rather than taking the time to plan everything out up front, teams can instead simply get started and use the opportunities to provide feedback that is presented in the time allotted for Sprint Review to adjust workflow as needed. This can be even extended to the creation of the Product Backlog. What's more, things could even go so far as for the Product Owner to emerge empirically from the available stock of stakeholders instead of being arbitrarily assigned.

While you and your team may initially find that starting a sprint without first completing a product backlog can be difficult, it is nevertheless, completely possible assuming of course that the team already has at least a general idea of what is required in the business in question and a project or project charter to work from. Using this information, your team should be able to, at the very least, come up with a realistic idea of what their first Sprint should consist of prior to getting started. Remember, the key words here are getting started. The goal during this initial Sprint is to be able to come up with something that can be demoed, and possible even shipped.

While sometimes this will result in things that are completely wrong or otherwise not shippable, the point isn't to get everything right straight out of the gate but to instead get everyone working through the Inspect and Adapt cycle as quickly as they can. This, in turn, will require actual stakeholders to attend the demo at the end of the initial Sprint in order for you to determine how successful it actually is.

Don't Worry About Advanced Tools: It is common for new teams to put off starting the Sprint process while they look through the myriad of different types of tools that are available to help them make using Scrum as simple and easy as possible. While there is nothing wrong with looking into finding a good Scrum assistance tool, it is important to wait to do so until you have a clear idea of just what facets of the Scrum process that you need electronic help with.

The impetus to do so is obvious, especially in teams that already work for technology companies. After all, why wouldn't you use technology to solve this problem as well as all of your others? In reality, however, Scrum electronic tools can often be more difficult to work with then their more analog equivalents. This occurs for several different reasons that can generally be broken into three different categories. First, they actually make it more difficult to freely share knowledge. Second, they actually can make the information that is shared through them successfully less clear while at the same time they make the information that is shared available more slowly to interested parties.

Looking into fancy tools this early in the process is akin to putting the cart before the horse, it is more important to get started at all then to get started in the perfect way with the perfect tools. Being extremely particular about a specific tool set is an easy way to put off starting to use the Scrum system while still feeling productive about doing so. Don't let yourself fall into this trap, get started with what you have available and work on improving from there.

When first getting started using the Scrum system, you should find that all of your tracking needs are easily met with a simple pen and paper tracking system. This will let you get started quickly and easily and get into your first Sprint without first ensuring that everything is as absolutely perfect as possible. Additionally, most of the time you will find that the simplest and most effective way of reliably improving communication within your team is by simply ensuring that everyone who is working on a given project is physically located near one another.

Additionally, you are going to want to ensure that the space has plenty of whiteboards or other writing spaces and that everyone is facing the center as opposed to being segregated off into their own little spaces. Finally, you are going to want to ensure that the group is separated from the rest of the organization so they feel as though they have a bit of privacy.

Keep your Product Owner Involved: A common problem that many Sprints face over time is a Product Owner who is energized and eager to contribute at the start of the Sprint, but who then falls off when the time comes to actually turn ideas into reality. A Product Owner is just as much a part of a Sprint team as anyone else which means they should ideally be present at every Daily Scrum as well as during the

Sprint Planning Meeting and the eventual Sprint Retrospective and Review. The Product Owner will also need to be available during work hours to provide input as needed to team members with questions. Essentially, when the Product Owner is not actively dealing with shareholders they should be involved in the Sprint Process.

Remember, the Scrum Team model was created with the express purpose of being as beneficial to productivity, creativity and flexibility as possible, but it can only do this when the Product Owner is as committed to respect, openness, focus, courage and commitment as the rest of the team. The only exception to this is if your team follows the strictest, most recent interpretation of Scrum which limits the Daily Scrum to the development team exclusively. There is still the option for others to observe the meeting, however, and this should include both the Product Owner as well as the Scrum Master.

While early on in your team's experience with the Sprint, it may be difficult to determine if your Product Owner is putting in enough time and effort, you will be able to tell if they added enough input, however, by how they respond to your Sprint Review. If during the Review, the Product Owner uses the time to provide feedback on the results then you will know they weren't involved in the process enough. The sign of an involved Product Owner is when they are leading the discussion with stakeholders, users or customers during the Sprint Review instead.

While the Product Owner needs to be as available as possible, there are numerous different scenarios where they may need to legitimately be absent from the process for a prolonged period of time. First and foremost, is when they are working with shareholders, though these types of meetings should be scheduled separately from Sprint based commitments in most cases. Regardless, it is going to be best for the Product Owner to leave a surrogate in place to provide their input for them based on the information that is already readily available.

Don't Use Stretch Goals: Stretch goals have long been used in many settings as a way of planning out additional goals that can be reached assuming certain conditions are met. They are antithetical to the Sprint methodology, however, and should never be used while on a Sprint for any reason.

There are two different types of stretch goals that can try and creep into your Sprints, it is important that you are aware of both of them and where they come from so that the Scrum Master can ensure they stay away from the positive team-centric attitude that the Sprint is cultivating. The worst type of stretch goal that can often appear to ruin group cohesiveness is the stretch goal that is suggested by someone from outside of the team.

Stretch goals that include a predetermined scope and deadline are the most difficult for Scrum teams to work through simply because it requires the team to start with a deadline and then work backwards to determine how to complete it, practically the opposite of how things normally go. What's worse, when objections are raised, they are typically met with either a reaffirmation to "get it done" or by simply adding more people to the team which does nothing to improve the state of things if the new individuals don't actually know anything about Scrum.

Otherwise, stretch goals have been known to pop up from time to time when members of the team try and determine what the Development Team is capable of, despite the Development Team's objections. It is important to keep in mind that the Development Team is going to have the best idea of what it can accomplish given various external limitations and wasting time disagreeing with them is, quite simply, time that can be better spent elsewhere.

The solution to this common issue is exceedingly simple, let the team determine the amount of work it can realistically get done within the confines of the Sprint. This doesn't mean that the first assessment of what is going to be done is going to be set in stone, but it does mean that no member of the team should ever feel pressured to commit to more work than they feel they can realistically complete. Forcing a team to commit to stretch goals is likely to build distrust among team members if the stretch goals are not met successfully. Distrust can lead to resentment and both will ultimately result in lower quality work being produced time and again. After all, if the stretch goal failed then there must ultimately be a reason why do your team a favor and avoid stretch goals and the witch hunts that they will always ultimately engender.

Make it Clear that Individual Sacrifice is Not Required: When your team is in the middle of a Sprint, they are going to routinely be forced to work through complex problems on the fly. There is a right way and a wrong way to solve these problems, however, and the lines regarding which is which can seem blurry without the right context. First and foremost, it is important to understand that the team naturally improves by coming together and solving these sorts of problems, which not only makes the unit as a whole more prepared for the future but more cohesive as a team as well. As such, if one person cracks the nut the entire team was working on then great, but if that person goes to near super human lengths to do so, then the team learns a very different lesson instead.

Rather than learning to work together to solve problems, the team in this case then learns to rely on the person who they know will always come through in a pinch. What's worse, this person could then start having an inordinate amount of pull over the team as a whole which means they could end up assigning mandates and even setting stretch goals without even realizing they are doing so. They can also cause several other types of issues for the team, starting with preventing the other members from developing an ability to creatively experiment and solve problems.

If one individual always has all of the answers then they are robbing others of learning why certain answers exist, which harms the team as a whole overall. If this proceeds unchecked for a prolonged period of time, then it is also likely that certain members of the team will become apathetic about the process as a whole because it will quickly become apparent that relying on the star is out of place with traditional Scrum practices. Ultimately this will break down the concept of the team at its base level and nothing will be accomplished, even if the star is still doing everything in their power to shine.

The easiest way to avoid this type of issue is to nip it in the bud during the planning for the Sprint. As long as things are planned properly from the beginning, there should be no need for any individuals to have to step up in any extreme way to get things done. Additionally, you may find it helpful to pad out your Sprint times if you find this type of scenario forming on a regular basis so that you can prevent things from devolving towards the need for such feats in the way they previously have before.

Chapter 8: Stories from the Trenches

Terminales Portuarios Peruanos: This is a company built around port and maritime services in Peru. Their IT group develops software for its own internal processes and operations. Traditionally it had worked along a predetermined release map that, increasingly, wasn't hitting the targets it needed to. With a team of 50 people, the goal was to deliver a new product at the end of each cycle but the process wasn't iterative and the teams were always held up at the end of development which made it difficult for them to meet their goals. IT group's ability to deliver new software every 30 days. The company chose to implement Scrum to scale by minimizing and removing cross-team dependencies and integration issues while elevating transparency.

In 2017 the company came up against a future deadline that it new it absolutely had to meet so it went about setting up a Scrum team to ensure the deadline was met without issue. Their initial phase of the implementation process started by bringing together stakeholders to align business objectives with user needs before aligning the results with the Product Backlog. The Product Owner also worked with the teams and used Impact Mapping and Story Mapping to help order and refine the Product Backlog. The Product Owner and the teams worked together to focus on creating integrated software.

The organization delivered products with a traditional project management model that prioritized schedule management and activity tracking to a model based on product delivery and daily progress. The first release was launched within one month and the product was in production completely by three months. The traditional model would have seen the first release within three months.

Vodafone: Vodafone, one of the largest mobile communication providers in the world, operates in more than 30 countries around the world and partners with outside providers of nearly 50 more. One of these partners, Vodafone Turkey, provides service to more than 20 million subscribers but the telecommunications industry in Turkey is extremely competitive and they needed help to manage the extreme Time to the Market pressure they were feeling.

Overall, there were three situations that needed to be dealt with in order for the pressure being put on the demand for improved productivity to decrease. The first was the time to market pressure, but there were also increased business expectations, an extremely long time-to-market period and a communication gap between all parts of the team. While the first issue could only be solved by improving the second and third, Scrum had answers for each.

The long period of time between when a product was developed and brought to market was due to the fact that testers and developers were considered to be separate units as opposed to one Development Team. As a result, the delay between the handoff for the two was significant and also decreased the responsiveness of IT as a whole. Likewise, the communication gap that the previous system created was solved by the added transparency that comes to the Scrum Framework and the understanding that everyone is working towards the same Sprint Goal at the same time.

To counter these issues the company set up a Scrum Team within the IT department in hopes of shortening their turnaround time while increasing the quality of the overall product as well. Under this pilot program, several Sprints were performed with this new Scrum team and the progress between each was tracked. Thanks to the added efficiency found in the Scrum framework, the pilot team ended up tripling its overall output in just three months.

Faced with such impressive results, the company decided to move forward with scaling the Scrum framework throughout the company. After about five months of getting things situated to the new way of doing things, the company reported that, across the board, the Scrum teams were performing at double the efficiency that was seen with the old model. What's more, the company also noted a marked decrease when it came to customer complaints as well as reported defects.

SoftwarePeople: All the way back in 2004, a company by the name of SoftwarePeople, based in Denmark, partnered with an investment firm in Bangladesh and made the decision to create a new subsidiary company in Bangladesh directly. Then, it hired 20 people in a single week and started using traditional processes in both offices with the end goal of receiving a CMMI level 3 certification in about 18 months. This led to nothing but countless pit stops and hurdles until the team switched back to Scrum in an effort to promote easier communication. The end result was the cessation of long-running projects, difficulties when it came to integration and technical issues and the adoption of smaller work batches, improved integration at all levels and a much faster delivery of clear business value.

It started in 2006 when the company was looking for a way to get away from their CMMI process that wasn't working. Someone within the company had heard about Scrum and everyone thought it sounded interesting so the CTO and three of the project managers decided to give it a shot and took a Scrum Master training course. That same month the CEOs from both countries got together to receive Product Owner training in the UK to ensure they had a shared understanding of what made Scrum unique.

Soon after they started experimenting with Scrum in the Danish office when a particular complex project appeared on the horizon Not only was the Scrum team able to complete the project with flying colors, their results were so compelling that they convinced the company to institute a rollout of the framework to both offices.

Their implementation strategy revolved around starting with Scrum teams working on customer and research and development projects so that they could develop a reliable rhythm. The process would then expand to global teams who would now be able to work together more easily due to the additional understanding Scrum brought to the project. After a number of Sprints, they saw positive results across the board and a 100 percent increase in efficiency in some cases. While some team members still had questions from time to time, everyone generally had a much clearer idea of what their responsibilities were and what tools they had to ensure they met their goals.

Conclusion

Thanks for making it through to the end of *Scrum: The Complete Step-By-Step Guide to Managing Product Development Using Agile Framework*, let's hope it was informative and able to provide you with all of the tools you need to achieve your goals.

Just because you've finished this book doesn't mean there is nothing left to learn on the topic, and expanding your horizons is the only way to find the mastery you seek.

The Scrum framework offers something for virtually every type of business, but it is important to understand that it takes time to start seeing even the most basic of results. As such, if you are preparing to be the flagbearer for Scrum at your company it is important that you understand that it will be quite some time before you will start seeing results as getting a Scrum team to work together effectively is all about training and practice. Even with the short-term hit to productivity, however, the end result will still prove far more effective which is why it is still a quality value proposition despite the required training. Remember, molding your team into a Scrum team is a marathon, not a sprint, which means slow and steady wins the race.

Finally, if you found this book useful in anyway, a review on Amazon is always appreciated!

Kaizen for Small Business Startup

How to Gain and Maintain a Competitive Edge by Applying the Kaizen Mindset to Your Startup Business and Management- Improve Performance, Communication & Productivity

Introduction

Kaizen in Japanese translates into improvement. In business terminology, it is used to refer to continuous improvement of functions that involve everyone—from the CEO to interns to assemble line employees. It can be applied to processes such as purchasing, supply chain, logistics, and more. Additionally, it encompasses multiple sectors such as life-coaching, banking, tech startups, and so on. The Kaizen philosophy is attributed to Japan's massive industrial glory after the devastation of World War II. It is said to be responsible for transforming the nation into a vibrant, economically flourishing, and cutting-edge economy.

The words 'kai' and 'zen' roughly translate to mean break apart and examine or improve/enhance an existing situation. It involves using both common sense and a meticulous scientific approach using quality control, dynamic framework of organizational principles, and beliefs that keep managers and workers focused on minimizing defects to optimize efficiency and productivity. Kaizen is not just another set of glorified quality control principles—but a way of life.

After the Japanese automotive and tech companies skyrocketed their way into the big league, Westerners began taking note of the Kaizen philosophy. Today, innumerable American and British companies have adopted the Kaizen philosophy to streamline their business, optimize profits, organize the system, and eliminate waste. In short, it is a quality management philosophy that combines the best of effective American business practices, quality management techniques, and *The Toyota Way*.

As a philosophy, Kaizen involves principles of improvement

based on commitment and cooperation. These principles are equally applicable in all areas of life—including your home, professional, and social life. The application in a startup, or any business for that matter, involves everyone from the top managers to the workers.

Typically, Kaizen systems involve total quality management, total productive maintenance, suggestion system, in-time production system, organization politics, conflict management, and small-group activities.

Here are some basic Kaizen principles:

- According to basic principles, people are the most vital assets in any organization. Teamwork offers results and presents everyone an opportunity to experience a feeling of accomplishment. It is believed that a dozen brains are better than one.

- Every worker must embrace continuous change and improvement. Ideas for employees, suppliers, customers, and manufacturers must be taken into account for discovering newer, better, easier, and more efficient ways to do things.

- Small changes or baby steps are easier to embrace and accept than comprehensive overhauls. Employees are more open to slow, gradual change than a sudden transformation. Small changes can also reveal that tiny changes can lead to huge positive results.

Conventional and established processes of methods of doing things can be comfortable. However, they may not necessarily be very effective. Everyone in the organization has to accept that change is necessary and good for the survival of an organization. An organization thrives on changing systems and processes.

Making excuses can lead to the downfall of a business or organization. Justifying old processes in the name of 'this has been working for the organization, and there's no reason to change it' isn't a constructive approach. Sticking to old ways can impede or hamper an organization's chances of being able to keep up with competition and changing times.

The Kaizen philosophy also says that if a job is done right the first time, waste is eliminated. Waste makes up for about 35 percent or more of the manufactured product. When product wastage is reduced, profits, and overall returns invariably increase.

If process errors aren't rectified immediately, they lead to bigger issues. Think about how equipment breakdowns or large scale failures are a result of letting a small issue go ignored and snowball into a major disaster.

By enhancing effective standardized programs, Kaizen attempts to remove waste. At its basic level, Kaizen is a business philosophy with a focus on endless and progressive momentum. Applying this philosophy to small businesses and startups can lead to plenty of improvements in various aspects of the enterprise such as employee morale, business efficiency, and overall profitability. This book talks about implementing powerful Kaizen strategies in your startup one step at a time to make its processes more organized for optimizing profits, employee loyalty, and reducing waste. Though it is primarily related to businesses, the Kaizen philosophy can also be applied outside business and productivity mechanisms.

By applying Kaizen principles to your startup, you will not just make the processes more efficient but also boost your profits, employee morale, and overall effectiveness as a new business.

It is important to understand that the concept of Kaizen is not limited to just one aspect or area of the business like marketing, production, distribution, and so on. It is based on the philosophy that improvements should be made anywhere they can be made without being limited to a specific area. While Western philosophy generally dictates that, "if it isn't broken, don't fix it."

On the other hand, Kaizen philosophy states that continuously strive to make it better improve upon it and enhance it—even if it isn't broken. If we don't keep improving and enhancing things, we are unable to compete with people who do it. Kaizen includes several well-known Japanese systems and processes such as process automation, quality control, suggestion systems, in-time delivery, Kanban, and so on. Kaizen is about setting standards and then continuously working on enhancing these standards to optimize results. According to the Japanese, there is always a better, faster, easier, and more efficient way of doing things. To support these lofty standards, Kaizen involves offering training materials, equipment, and supervision that is required for employees to accomplish higher standards and keep up their ability to fulfill those standards on a continuous basis.

Here's how the Kaizen activity cycle can be defined:

1. Standardize activities and operation

2. Measure standardized activities and operation (identify cycle time and in-process activity.

3. Evaluate measurements against needs/objectives.

4. Innovate to fulfill requirements and enhance productivity.

5. Standardize improved and enhanced operations.

Keep the cycle going. The most crucial Kaizen elements are the effort, involvement of every employee, quality, the eagerness to keep changing and improving, and communication.

Some features of Kaizen are particularly noteworthy. The Kaizen philosophy focuses on processing enhancements and improvements rather than arriving at a specific target result. It needs consistent and ongoing management initiatives to focus on improvements. Ideally, this should comprise 50 percent management efforts. Kaizen is about incremental improvement ideas rather than breakthrough inventions. Small changes around continuous incremental ideas are the key to successful implementation of Kaizen.

Several Kaizen ideas can be implemented on very small levels to make processes more streamlined and efficient—while also ensuring employee satisfaction. Employee satisfaction is a huge aspect of the Kaizen philosophy. It places worker's satisfaction as a top priority for bringing about improvement and enhancement in processes. An improved way of doing something will typically be adopted if they make things easier, more efficient, and more fulfilling for workers. Though Japanese workplaces are characterized by a well-defined process and work standard set, they are constantly evolving over a period of time. The employers are more than willing to respond to improvements suggestions, and even set systems in place to track the progress of these suggestions over time.

Japanese Kaizen philosophy fundamentally believes that:

- There is no end to the process of improvement. It is a continuous and on-going pursuit. You can keep striving for perfection but you can never say, "Everything is perfect, and nothing needs to be changed now." Change and improvement is a continuous and ongoing process.

- There must be a great desire, mindset, and curiosity for conceiving improvements. Unless you have an experimental, explorative, and curious mind—you will not attempt to come up with different ways to do things. Even the tiniest improvement can make a huge difference. Big changes made once in a while are not as much Kaizen as small changes made consistently and continuously over a period of time.

- Improvements can originate from people who are consciously looking for results and details. When you are detail oriented, you know exactly what works and what doesn't. Train your managers and supervisors to figure out different methods of improvement, while also improving upon improvements.

Chapter One: Kaizen and Teamwork

Kaizen is at the epicenter of teamwork and team-building strategies. While teamwork has become the magic word of today's competitive business circles, very few organizations make a sincere effort to streamline teamwork to make their systems function more efficiently. This reveals a major disconnect in teamwork theories and practical implementation of the philosophy of teambuilding.

As part of the Kaizen philosophy, every worker is encouraged to offer his/her ideas and suggestions for improvement. Even the tiniest changes and suggestions will be considered and implemented. For instance, Toyota team members (note they are team members and *not* employees) are motivated to identify minor details that can use enhancement processes to make the desired improvement. Think of an unorganized file system that can be streamlined by alphabetizing and color coding.

While several organizations do not permit file purging as part of the process in office, it can be tons of paperwork, outdated, obsolete, and creates physical as well as mental clutter—this can be time, energy, and efficiency-sapping. Then again, Kaizen can be used for better time management skills. You can go through the clutter and eliminate what no longer serves a purpose. Kaizen principles can involve giving employees time for going through paperwork for cleaning out their desks, workstations, and cabins to begin afresh for maximum productivity. This isn't related to teamwork but nonetheless an important aspect of Kaizen.

A Kaizen team engagement event in a business or organization involves improvements and team bonding through teambuilding exercises. People pick up new skills and new ways of looking at their organization and functions from a system based perspective. It urges employees to take responsibility or ownership for their contribution to the entire process as well as for pitching in to the overall improvement of the enterprise.

As a startup, every employee or team-member is made to feel responsible or to contribute to processes that add to the overall improvement of an organization. There is no single person or leader responsible alone for the driving process as well as system and operational changes. It is *every* employee's responsibility to contribute to the success of overall business processes.

In Japanese organizations like Canon and Toyota, employees usually make 60 to 70 suggestions annually. These are not just heard but also documented, sent out, and implemented. In majority of cases, these are not major suggestions or ideas. They are based on making small changes on a *periodic* basis to help bring about change and improvement as well as to make employees feel a part of driving that change as if it is their own company. This fosters feelings of teamwork and loyalty, which, in turn, helps reduce employee attrition and boosts an organization's overall productivity.

There are different phases of team formation, growth, and development of a startup. The first is when a new set of people are just getting to know each other. They are all wondering not just about their purpose but also about processes and systems they've been pulled together to facilitate. Then, the team starts performing or working together.

In this nascent stage, a good facilitator can help review the team's performance and identify the current performance and areas of improvement. Think about it—always have a diverse group of people throwing in suggestions. It should be a mix of people who are actually involved in the process, people who wholeheartedly support the process, and people who are completely outside the process. This helps us gain a fresh perspective on the way things are done—while also bringing to light more wasteful pursuits.

Chapter Two: Implementing Kaizen in a Startup

There are plenty of ways to implement Kaizen in your startup to gain a competitive edge right at the beginning now that you know what the philosophy is all about.

Every startup founder dreams of making it big with his product or service. However, not everyone can fulfill their dream. Only in every ten thousand startups can skyrocket their way to success and growth. This isn't mere destiny or a matter of chance. It a result of carefully made decisions and an atmosphere that breeds constant improvement and productivity! Results happen only when there is an ideal environment for growth and development.

Training

At the onset, startups can ensure that employees are trained for future changes, dynamics, and advances to retain an edge over other business. Another important aspect is to train current employees to train future employees. A training culture breeds improvement and continuous progress/enhancement in any organization.

It is the key to creating an environment that leads to open sharing of innovations, suggestions, and ideas. Every employee should be made to feel like a stakeholder in the company, which is when they will feel a part of the procedural and decision-making process. Train employees not to follow orders—but more importantly, to be decision-makers—if you want your startup to evolve into an organizational force to reckon with.

Team Building

Like we discussed in the earlier chapter, building solid teams is at the core of the Japanese Kaizen philosophy. Going by the Kaizen tradition, your team should include both experts and employees who are willing to challenge power positions. There has to a change in the status quo if you want to strive for transformations in the system. Sticking to a rigid status quo goes against the Kaizen philosophy. Change and continuous innovation are key to successfully implement of the Kaizen philosophy.

Developing Improvement or Driving Enhancement

How does one develop or drive continuous improvement or enhancement in a startup? Think brainstorming, ideating, keeping an open door policy for suggestions, creating a less rigid organizational hierarchy, and so on. This is where data collection comes into the picture. Start collecting data about your business from day one. Use it to understand what is working and not working in the startup's favor. Though it is fairly early to drop systems that are ineffective at the beginning where startups are concerned, a collection of data over a period of time can help you make fairly accurate decisions and drop ineffective systems in the long run. As an entrepreneur or founder of the startup, you may not use these records. However, they are important for decision making and dropping ineffective systems in the long run. Collect data, analyze it and work out what changes need to be made in the long run for the efficient functioning of the startup.

Implementing Change

The biggest enemy of progress is an organization's unwillingness to change its procedures, policies, and systems to embrace newer, quicker, and more efficient ways of doing things. Assess the effect of these changes on the progress of your enterprise. Start by assessing the impact of changes over a period of a month. Then, switch over to assessing the impact of procedural, system or other changes over a period of one year to ensure the changes are effective.

How Do You Implement the Kaizen Philosophy Within the Context of a Startup?

At its core, Kaizen is all about teamwork, which means you have to involve all your workers completely involved in the systems and processes. They should be encouraged to share their ideas, analysis, observations, and suggestions about the business as if it is their own. The ideas and suggestions should involve optimizing business processes, increasing work productivity, saving resources and improving quality/safety parameters.

Employees must get the required stimulation for actively participating in and realizing their observations and ideas. Startups may not have the budget to hire a Kaizen expert. You can overcome this by attending Kaizen trainings and seminars to understand the finer nuances of this progressive Japanese philosophy and start applying it to fulfill your organizational/business goals.

Employees should be encouraged to gain a six sigma certification. Then there are other trainings such as project management, which can help improve their overall efficiency by minimizing errors and streamlining the system.

The Do's and Don'ts of Applying Kaizen in a Startup

Kaizen goes beyond simply getting employees involved in the

process of improvement and development. Here are some dos and don'ts of the Kaizen philosophy when it comes to implementing it in your startup.

1. Avoid relying only on your expertise or have excessive faith in your abilities alone over others. Consider of the skills, ideas, wisdom, and experience of other people in the team if you want to evolve in a competitive and progressive business. Following rigid systems and procedures can impede your results, thus allowing other more change embracing competitors to move ahead in the game.

2. Do not blame others. Accept responsibility for your actions and decisions.

3. Replace "We can't" or "I can't" with "we can" or "I can."

4. Treat other people as you'd want them to treat you.

5. Do not wait for accomplishing perfection. You can begin a business even with 50 percent improvement, and keep improving and making changes along the way. If you wait for until everything is perfect, you'll only be led into a state of inertia.

6. Avoid overlooking problems and issues. Rectify them as soon as you notice it instead of delaying it or waiting for the right time to resolve it. The right time is when you discover. The more you overlook small, everyday niggling problems in your startup, the bigger they grow into and cause heavy disruption in the overall productivity, efficiency, and success of the startup.

7. Avoid hesitation when it comes to asking someone (even an intern or a new employee) about something that you

don't understand. When things are not clear, do not allow the ambiguity and vagueness to build. It is better to clarify things in the beginning than allow misconceptions and an unclear understanding of systems, procedures, and technology to build over a period of time.

8. One of the biggest mistakes many startups make is following opinions over statistics and figures. Since it is a new business, everyone desires to offer their two cents. However, this isn't really productive where efficiency and results are concerned. By all means, embrace all suggestions and ideas. However, ensuring where ever possible that they are backed by statistics. Keep in mind that improvements are not made over cups of coffee in the conference room, which means 20 percent discussions and 80 percent practical tasks.

9. Never believe that you have everything in perfect order and that nothing needs to be changed. You need to strive for continuous enhancement, change, and improvement. This is especially true in a startup where you will be testing different system, organizational and procedural models to know what works best for you.

Tips for Developing a Kaizen Mindset

Kaizen has value for both—the growth of your startup as well as you as an individual. Though in this context we are talking about applying Kaizen to your business, it can well be used in both your personal and professional life. Making it a way of life has allowed the Japanese to lead a fulfilled, rewarding and gratifying life. Here are four basic steps for acquiring a Kaizen mind.

1. Plan.

Plan your day on the evening of the previous day. Know what you are going to focus on the night before. Once you have decided the focus of your day, plan your other activities around it. When you know what is lined up for the day before beginning the day, it becomes more streamlined. You attend to high priority tasks that need your time and attention first. Similarly, you don't waste time and energy trying to figure out what to do. Everything that you need to get started with your work is already in place. All you have to do is begin.

2. Visualize.

Visualize your day by waking up early each morning. Make mental notes or allow the mind's eye to see how you plan to live the day. Think of everything that you plan to do throughout the day in a step by step manner. These mental footnotes can successfully place triggers within your subconscious mind, which can guide you in the right direction. The thing about our subconscious mind is that it doesn't know the difference between reality and imagined reality. It believes whatever we feed into it to be real, thus guiding our actions in line with what we visualize (and what it takes to be real).

3. Set Reminders.

Setting reminders is crucial to adopting a Kaizen lifestyle. As a startup, you'll have to do plenty of things to set the system in place. Remember, it is when the baby is just born that it needs maximum care. Once it is up, walking and running on its own you simply have to keep a watch but you don't have to carry it everywhere. Simply, a new business involves setting multiple policies, procedures, and systems in place. And if like me you don't have an elephant's memory, you may not be able to complete all tasks on time to strive for improvement. Smartphones are great for setting reminders and schedules. Ensure you stay on the path of continuous and consistent improvement by setting reminders and sticking to a schedule.

4. Review.

Towards the end of the day, take notes about how things went even if you cannot recollect the exact goal. The lessons that you acquired through today's review can be utilized for creating tomorrow's plan. Reviewing is absolutely integral to the Kaizen philosophy because it involves continuously learning and making improvements based on past experiences and performances.

Keep in mind that Kaizen doesn't ensure immediate success. It is a slow and gradual process. However, results are certain if it is used effectively to drive continuous improvements and enhancements. Here are a few areas where Kaizen can be used for accomplishing amazing results.

Five Areas Where Your Startup Can Benefit by Employing the Kaizen Philosophy

Kaizen can be as simple or as complicated as you want it to be. At its heart, it's a simple and effective philosophy that has the potential to bring big results. However, where most businesses struggle is knowing areas where Kaizen can be implemented or identifying methods that can be used a part of adopting the Kaizen philosophy.

1. Management Performance

As a founder or manager of the startup, you need to be convinced about the Kaizen philosophy before implementing it across the organization. This can be done by designating a time

(15-20 minutes each day) for every manager or key personnel to devote time to Kaizen related activities or work. It can be anything from organizing their workstation to labeling their files to keeping the desk free from clutter. Make it a policy for every manager to spend a few minutes of their workday practicing Kaizen. Train your managers and key players to keep asking why. Go to the root of the issue. What is causing a certain problem? Why is something happening? Managers can use this strategy to recognize what is happening and when an issue arises.

Then again, managers in the startup should be trained to work as a unified team for containing and correcting the root cause of a problem. This involves strengthening interaction and communication between the management team or managers and employees. As a founder, you will have to carry everyone along on your way to success. This ensures that key players are always aware of issues, and the team can be involved in identifying solutions.

2. Management-Employee Communication

Kaizen is about enhancements in the way managers and employees interact or communicate with each other. It fosters the idea that management and employees are partners or stakeholders in the startup. The rigid hierarchical system that prevents employees from communicating with managers or founders in a startup doesn't foster progress, improvement, and development. Everyone should be viewed as working together with each other rather, thus removing the 'boss-servant' equation. It makes sense for the management to talk to employees as equals instead of dictators. It helps everyone recognize and encourage better ways of doing something. Try to develop a positive relationship with your employees, while attempting to understand what they do on a daily basis.

It can be something as simple as greeting co-workers and employees first thing in the morning and inquire about them. These small steps strengthen the founder-employee interaction and help the management be attuned with what is happening at the grass root level of their business.

3. Productivity, Systems, and Processes

When you improve your business' management/founder interaction, it leads to greater productivity or a strong tradition of process Kaizen, which creates improvements that can be implemented on the very same day. Primarily, this is a review of activities to determine if one can work more effectively or better. Start with a plan-do-check-act cycle if you want to incorporate Kaizen into your management process immediately.

1. Planning involves identifying opportunities and having a solid plan for change in place. This is where the staff and management closely work together to identify potential action or improvement areas.

2. Doing involves implementing small-scale trial changes to gauge if the change is worth it. For instance, you may notice that outsourcing your accounts and billing is helping you save time, and focus more on your areas of strength, you may try this for a while before implementing it on a full-fledged scale.

3. Checking is essential for reviewing systems and processes, where you seek to understand and analyze lessons and results acquired from the review.

4. Act on what you've learned as a result of the review and analysis. If the change is successful, it can be included in the system. If it isn't successful, start all over again with a different approach and focus.

4. Employee Performance

Kaizen implies every input is valued. No suggestion is too small or stupid to be implemented if it helps the overall productivity, efficiency, and bottom line of a company. This includes everyone from seemingly lowly cleaners to supply chain professionals to janitors. Everyone in the startup has a voice, which must be recognized and encouraged. You never know which suggestion or idea will be the game changer. It is also a given that if you want your workers to do well and produce stunning results, they must find their job interesting and satisfying. Naturally, one needs to increase their awareness of the fact that their thoughts ate highly valued. As a startup founder, you will also have to ditch the blame culture and foster an atmosphere where everyone is helpful towards one and another.

Create an atmosphere where everyone's opinion is respected. In fact, go a step ahead and proactively ask your employees and workers for suggestions and reviews. Think about it—your cleaner or janitor may come up with a more organized and efficient cleaning system that can help you save money.

Designate an open day, wherein any of your employees can walk into your office and throw a bunch of meaningful and relevant suggestions. This can be accomplished through a streamlined suggestion system that can be work effectively for small and large groups based on what works for the organization, and the way expertise is demarcated or divided within the startup. Everyone who works for the startup should be aware of their expertise and niche. They should be trained and encouraged to identify systems or procedures that can be changed.

5. Customer Experience

The Kaizen philosophy's ultimate goal is to enhance overall customer satisfaction. All other values and attributes of Kaizen are directed towards creating rewarding customer experiences. Ineffective business practices can lead to potentially dissatisfied customers, thus preventing your startup from growing in the long run. As a new business, you need to establish your customer base. Hence, simplifying customer experience and keeping them happy is integral to the Kaizen philosophy.

For accomplishing superior customer service, the startup must may every business procedure that includes the customer right down to the details. As a customer-centric organization, this helps you determine where complexities, complications, and challenges may arise. For instance, if you are an e-commerce website, you may want to check if the online ordering and payment checkout system is efficient enough for your customers. Do your purchase system and website facilitate a smooth, quick, and easy way to order? What are the small ways in which you can improve the website to make the processes of buying even easier for the customer? Maybe add other products that are similar to the one he/she is considering buying and compare it with their currently chosen product to offer them more options and help them make more efficient buying decisions? Maybe show them lists of products similar to what they've purchased that have been recently purchased by customers or which are popular with buyers. There are innumerable ways to improve online buying and payment methods. As a startup, you keep identifying ways to make the process of buying easier for customers.

Once you identify potential issues that can impede the overall customer service experience, you can start recreating or improving the process by using guidelines to rectify the issue. Follow-up with a test and review process, where you can gauge or measure if the changes have helped in making the process of customer or user experience amazing.

Now that we know in brief how Kaizen can be applied to a startup, let us look at the 5 S's of this cutting-edge Japanese philosophy.

Chapter Three: The Five S's of Kaizen

5S is the foundation of Japanese Kaizen philosophy. Though entrepreneurs and business leaders from the West have lent it their own twist, at its heart, the Japanese model of Kaizen philosophy is primarily driven by the 5S.

The Kaizen 5S model emphasizes on having visual order, neatness, organization, and standardization. Some results one can expect from the 5S model include greater overall efficiency, enhanced profitability, better safety norms, and more professional service standards. These are all things a startup strives to accomplish in the long run.

As a startup, your enterprise can experience several benefits of implementing the 5S program. On the face, it appears as though only manufacturing businesses can benefit from Kaizen. However, this is only a misconception. It can be applied to just about any sector—from professional services to healthcare. Every area or department within a startup or organization can benefit from the 5S model.

Here are the Kaizen 5s:

1. Sort

This is known as *seiri* in Japanese and sorting involves eliminating all unnecessary and defunct items from your workplace. These items are believed to hinder productivity, space, time and energy. Therefore, anything that doesn't serve a purpose in the functioning of your startup is best eliminated to make way for newer, fresher and more productive items that serve a clear purpose. Decluttering is a big component of

Kaizen. If you keep items you don't need, you'll spend plenty of time and energy looking for items that are required, thus losing precious resources in the bargain.

Seiri involves sorting out. As per *seiri*, employees should always sort and organize things. Label, classify or categorize items as, "necessary" "critical" "most important" "urgent" "not urgent" "most important" "not needed now" "useless" and more. The categories can be classified as, "urgent but not important," "urgent and important," "neither urgent nor important," and "important but not urgent." This way you know what tasks or files are a high priority and what can be tackled later. Throw away everything that is useless or doesn't fit into your scheme of things. You may be carrying things from a previous business to the startup. It may no longer serve a purpose to what you are currently doing. Ditch things that are no longer, valid, applicable or relevant to the current startup

2. Set in Order

This is also known as *seiton*. It speaks about creating a particular order and location for every item. Everything should be in its place if you want to be able to save time, money and energy in looking for it. Kaizen is about orderliness. Having a set order and system in place for things allows the Japanese to be more efficient. It leaves more time and energy to focus on procedural systems. *Seiton* is about organizing. According to research, employees waste more than half of their precious time looking for important documents. When every item is kept has its own place right from the onset, it is easier to find it when it is urgently needed.

Why is *seiton* important? It ensures systematic arrangement for effective, productive and efficient retrieval. You must be able to find where everything is kept, which means everything should be in its place for facilitating east retrieval. The place for every item should be clearly marked and labeled. There has to be a demarcation of space of each item right when the business is launching. When *seiton* becomes a habit, it ensures a smooth and unhindered workflow. Ensure that equipment which is frequently used is easily accessible. Similarly, employees shouldn't have to struggle too hard to access the material. This saves their time and effort, which can be productively channelized towards other systems and processes.

3. Shine

Shine or *seiso* is about cleanliness. Keeping your work area clean reflects a healthy, positive and productive state of mind that leads you to be even more efficient and productive. Will an ill-kept and dirty space inspire you to do good work or be productive? Little chances! *Seiso* involves organizing the workplace. A shining and clean workplace breeds good vibes that encourage positivity and productivity.

As a startup, you need to ensure that rules for cleanliness, neatness, and hygiene are set at the onset. Keep all documents in efficiently labeled files and dedicated folders. All bills should go in one file, while all invoices should go into another. Keep all checks in one file, while receipts can go into another. Simply following these small practices can help you streamline the process to make it more efficient. Try to come up with new ways to keep your space clean, shinning and organized. Use space optimally for storage. Ensure that there are enough efficiently designed and optimized storage options such as cabinets and drawers for storing your stuff.

Shining or *seiso* is an important aspect of Kaizen. Cleaning the workplace along with maintaining all the equipment regularly should be a part of your cleanliness and maintenance policy.

Keep your space clean, organized and tidy. After each cleaning session, follow up on it to ensure the cleanliness is maintained. According to the Japanese philosophy, a shinning workplace leads to higher efficiency and gains.

4. Standardization

Standardization or *seiketsu* is adopting best practices within the workplace by fixing a standard norm for systems and procedures. Each organization should have a set of standard rules and policies to drive its systems to ensure quality and optimize productivity. Work practices within the startup should be consistent, uniform and standardized to ensure greater discipline and efficiency. All workstations should be uniform. Employees should be able to walk in to their workstation every morning and view the same equipment or tools, which are kept in the same location or spot every day. Everyone should be aware of their responsibilities and commit to following the standardized code of conduct set for all employees.

5. Sustain

Sustain or *shitsuke* is continuously improving and never slipping back into old ways. It is related to self-discipline. All employees need to respect an organization's rules, policies, and regulations. For instance, not walking into the office wearing casuals (if there is a dress code) on a weekday or entering office without an identity card.

Most startups make the mistake of taking things lightly in the beginning, focusing on luring employees with the fun, light work atmosphere, and play syndrome. However, self-discipline is important for the long-term sustenance of a business. You may find a relaxed policy effective in the beginning. Once the business decides to scale up though or grow by leaps and bounds, it will be tough to keep up with a set of employees who aren't disciplined. They can end up lowering the organization's overall productivity, efficiency, morale, and atmosphere.

Once all the earlier 4S's have been set, you have to ensure that your organization doesn't slip back to its old ways by following *shitsuke* to the fullest. When you follow the earlier attributes, it becomes a way of life. Keep your vision focused on this new way of doing business, while not allowing yourself or your startup employees to slip back to their old ways. Remember, continuous and not momentary improvement is the key. Adopting continuous improvement leads to lesser wastage of time, higher productivity, better quality and rapid lead times.

There are several benefits of incorporating a culture of 5S's within the organization including higher productivity, reduced clinical defects, quicker, and more efficient products/service and a safer work environment.

While the first 3 S's are relatively simple, several startups face problems in the 4 and 5 S. However, it isn't something that is tough to accomplish. It can be achieved with consistent training, practice, and implementation.

Chapter Four: A Step-by-Step Kaizen Guide for Startups and Small Businesses

Identify Significant Business Processes.

Begin your Kaizen quest by identifying important business processes. While conventionally, Kaizen processes, much like Six Sigma, are applied to the manufacturing sector—it is now used in every type of trade and industry. Identify processes that can be used for your startup on a daily, weekly, and monthly basis. This can help you work on improving the processes on a continual basis. It is important to have emergency funds ready for effectively managing the cash flow. Keep a business credit card handy to manage emergency cash crunch situations.

If the processes aren't in place already (which is the case in a majority of startups), it is the right time to put documentation in place for every aspect of the business. Avoid relying on the expertise of a single person. One of the biggest mistakes startups make is relying excessively on the expertise of the founders—and not document processes. This isn't a good Kaizen practice. A good Kaizen practice involves documenting *all* processes for continuity and offering an opportunity to the entire team for reviewing procedures. There will be opportunities for better results or the creation of more efficient processes. You may identify a few old processes that can be dropped, modified, or replaced—leading to huge savings.

What are the businesses that move or drive your organization ahead or are absolutely important for the functioning of the organization? Carry our constant training and reviews, and use the feedback for improvement. When processes are documented, it is easier to find ways that can be improved upon. Your startup will grow and there will be massive changes in the business environment as the startup grows. You'll have to keep up with these changes, which is why adapting to different systems and procedures will be vital to the success of your startup.

Ask yourself these questions to get your startups Kaizen in place. What are you doing currently? How are you doing things? Is it possible to enhance what you are currently doing even if it is slightly? Maybe change the way you do your invoices so you don't spend a few hours each day handling them and tackle them all together on a Friday or Monday to make the invoice paying system more streamlined. There are several such questions you can ask yourself that point to the correct direction when it comes to implementing the Kaizen philosophy for improved efficiency.

Empower the Workforce to Make Continuous Improvements.

Is your organization a safe and conducive place for workers who offer suggestions for improvement or constructive criticism? If this isn't the case yet, your startup needs a Kaizen approach. As a startup, it is even more important for people to feel empowered, and bring about exciting new ideas on board. This will make them feel part of the launch of the business or startup, and increase their overall loyalty and morale. Think about it, each time an employee within your newly launched business is encouraged to come up with a suggestion, and their suggestion is implemented, they feel important, valued and empowered, which drives them to come up with even more suggestions for improvement and enhancement.

Give your workers the independence to fix or resolve issues

they come across. Of course, there should be a boss or experienced person helming the functioning and decision making within the organization. However, they should have the creative freedom to fix things and improve procedures to make their assignments more effective. If these processes are found, they should be shared with the team so that oo

However, it is also important for people to believe that they have the freedom to make fix things wherever they find issues. Encourage and motivate your workers to come up with innovative, creative and resourceful solutions to make their work procedures and assignments even more effective. Inspire them to come up with better ways to do things, and share them with others. Celebrate small victories to develop a culture of continuous and consistent improvement.

All employees in your startup should feel empowered to make decisions or play an active role in the implementation of decisions. This is vital for small businesses because you are in a startup launch stage where each employee's role is crucial. They are all a part of the founding or early team, which makes them feel a vital part of the organization's success, history, and story. When you empower these already important workers to make decisions, you make them feel even more strongly about their role in the organization. In a small business or startup, it isn't uncommon for a single person to represent the entire department, which is why open communication is vital.

Take suggestions, feedback, and reviews seriously whenever an employee talks to you. They may offer you insights from the ground level about something that has probably missed your eye. Your employees are the ones who are working hands-on on various processes and systems so they know best. Encourage them to talk about what is working great, and what isn't. Understand that they are experts in their particular processes, and they are able to closely examine how these processes function. This is exactly why the information sought from them should be valued and sought after.

When you implement the Kaizen philosophy within your startup, understand that no detail is too big or small. Everything is important and has value. Every opportunity where you can view even the tiniest inkling of improvement should be valued. Tiny improvements can add points that will end up becoming large improvements over a period of time. This is applicable throughout the company, and not just expenses. Waste can be discovered in several forms. Anything that is consuming an employee's time or even physical/virtual storage space should be studied and reviewed as a possible Kaizen application area.

Understand that improvement or enhancement is not the final destination of your journey. It is not the finishing line but instead a philosophy that one must continue their pursuit of exploration and improvement arising as a result of the exploration. Kaizen isn't about reaching anywhere. It is striving towards continuous improvement, one step at a time.

Strategies for continuous learning and improvement a must be keenly intertwined into your company's fabric and be a part of its basic culture. Don't restrict Kaizen implementation only to specific processes or occasions. It should be assimilated into your startup's organizational philosophy as a way of life.

Foster a "Yes Culture" by Incorporating a Win-Win Attitude.

If you've worked with a "no" culture organization in the past, it can be frustrating to work in an atmosphere where you are told something isn't possible, even when you know it is. You may have used the exact system, procedure or technology in the past only to be told that it is not possible to use it. Would you like to foster such an atmosphere of negativity within your startup?

Kill the "no" culture if you want your business to grow and thrive in the long run. Train your employees to say "yes" whenever they can. It can be for everything from embracing new ideas to helping a business grow and succeed in the long run. A "yes" culture is based on the philosophy that improvements, changes and helping other people succeed go together.

Don't Overlook Details.

Use the Pareto Principle or the 80/20 rule to dramatically enhance or improve your results. It is a good strategy to get the best results by making small efforts. Small business and startup founders should encourage their employees to seek big wins where the startup can gain the biggest efforts with the smallest and quickest possible efforts. However, this doesn't mean you should ignore the smaller details. Remember, details can make or break your business. A spectrum of industries ranging from software development to manufacturing to those operating within the service sector can benefit big time from making small changes or fixing small details. For instance, a tiny change in the manufacturing process can cuts wastage by 4 percent can translate into a 4 percent decrease in raw material cost. Over a period of time, this 4 percent can add up to a huge amount.

Make Continuous and Consistent Improvement a Part of the Organization's Culture.

In certain organizations, improvements only happen at fixed times during the year. Let your startup not follow a similar culture where changes and enhancements are only encouraged during specific times of the year. Don't restrict improvements, process changes and system enhancements to certain times of the year or annual meetings. Make it a part of your daily organizational culture. Build your startup an improvement driven entity right from the beginning.

Your workers should be excited about coming to work every day. They should be driven by the goals, values, and philosophy of improvement to keep making changes on a continual basis within the organization. Employees will be enthusiastic about coming to work each day when they find a more efficient way of doing things, and they realize that they are free to implement this effective way of doing things. They should be encouraged to add to small victories and be given the freedom to make decisions based on what they feel is right for the organization's overall quality control, development, productivity, and profitability.

At times, even the most unexpected workers may surprise with their massive gains that can enhance your company profits big time. When this happens, it's a true win-win scenario for any organization, which is exactly what Kaizen is all about. There are no easy fixes in Kaizen. The simple yet cutting-edge Japanese philosophy doesn't have a magic bullet that will eliminate your problems or make challenges disappear in a shot. Since Kaizen means "change for the better," its simplicity makes it special because is about constant improvement, even if it is a slight improvement.

Build Your Startup Around the Kaizen Spirit.

Everyone loves stories of quick and radical changes that produce instant results. However, a more doable and effective approach involves bringing about organizational and individual changes through a series of tiny and systematic steps.

The way it works on a psychological level is that radical changes set off our brain's fear and response system, thus shutting our capacity to think in a clear and creative manner. On the other hand, tinier and quieter steps prevent stress, anxiety, and fear. It doesn't trigger our brain's alarm or reflexive defense mechanism. Our creative, logical and cognitive processes flow easily when there is a lasting and power-packed change in a gradual and incremental manner. This is why the adage "slow and steady wins the race" holds weight.

Similarly, constant repetition makes it simpler to transform activities into habits and eventually lifestyle. The act of doing tiny tasks gives your mind a strong sense of accomplishment. Keep making small and gradual changes to processes and systems on a daily basis instead of making radical innovations that can scare you. When you make small and gradual efforts, your mental blindfolds are overcome and work-related anxiety takes over to prevent you from wholeheartedly embracing change. Gut-wrenching change can overwhelm the best of people. Think about major changes and the resulting challenges of the changes faced by organizations by Ford and NASA.

Start applying these actionable Kaizen pointers in your startup right away to witness big results in small ways:

1. Product and Service Innovation

One of the best ways to keep improving upon your product or service is to stay curious and observe the tiniest of details. This practice alone will offer you a huge edge over competitors. Learn from frustration, pain points, disappointments and failures instead of letting them get the better of you. Learn to be curious, keep learning and ask questions all the time. Cross collaborate with startups that are similar to yours.

2. Quality Improvement

Always be on the watch out for warning signs that allows for confidential and secure reporting of mistakes. Every process must be tracked, even something that is working wonderfully well in favor of the organization. Discuss systems and procedures, and don't duck bad news. View it as an opportunity for change and reinvention. Overconfidence is shunned by the Kaizen philosophy. One must never assume that they know everything or that there is nothing left for them to learn. Improvement and change should be continuous. Pixar is open to suggestions from everyone from the CEO to the newest employee on board.

3. Cost Cutting

Invite employees for suggestions on taking small steps to slash costs, while being open and receptive to a bunch of well-meaning and relevant suggestions. Offer small yet meaningful rewards as positive reinforcement for cost-cutting suggestions. It will be worth it if you manage to hit upon that one idea or suggestion that can help you cut costs dramatically. For instance, Continental Airlines offers its employee a $65 bonus for higher performance, American Airlines solicited suggestions to save a minimum of $25 and so on.

4. Boosting Employee Morale and Overall Job Satisfaction

A happy employee translates into increased productivity and higher profits. Create small gestures of demonstrating respect and gratitude towards your employees. Make them feel a valuable part of the organization. Show respect, interest, and curiosity towards what they are doing. Always, always greet your employees with their name. Pay careful and mindful attention during conversations and interactions. Attempt to find out more about the fears, insecurities, dreams, goals, objectives, challenges and aspirations of your employees by asking more open-ended questions! Ensure you communicate with them on a regular basis. Invite them over for a chat in your office and discuss how things are going or if they have any suggestions for improvement.

Did you know that Zappos hires only people with a highly positive attitude? Hell, they even ask their drivers for their views about candidates they picked and dropped to gain insights about how positive they are. Also, medical clinics are known to ask receptionists for ideas about improving overall customer service.

5. Increasing Sales

How can Kaizen be used for driving sales and making the process of increasing profits more effective? Motivate your sales staff through guided imagery, visualization, vision boards, and mind sculpturing. Keep reminding them and yourself about the organization's mission and values. Volvo has a highly effective practice for boosting employee morale. It circulates letters from satisfied customers to employees to make them feel an important part of customer satisfaction process.

Innovation in large organizations and startups is driven by problem-solving, creativity, break-through thinking, problem-solving, inspiration, discipline and much more. If you think about it, inspiration is much likelier to develop and strike through the habit of continuously paying attention to and observing your venture's tiniest moments. Contrary to popular perception, creativity isn't some flash of a lightning moment. It is a result of careful and conscious activity. Silly mistakes, trivial issues, and even boredom can lead to innovative and new ideas. Train your employees to respond to their curiosity. Ask more open-ended questions, and be patient in your quest for answers.

For instance, James Fargo of American Express was simply frustrated with cash and card processes while traveling. This led to the invention of the American Express Travelers' Check. Train your employees to use their boredom and frustration for channelizing change.

Similarly, UK based scientist Shashikant Phadnis discovered a sweetener by accident when he was asked to "test" a chemical, which he misheard as "taste." Mistakes and accidents are a part of the organization's evolution process.

Kaizen is the universal desire of workers to feel like an integral part of an organization and engaged. Whether it is related to innovation, driving results or productivity, key personnel, and startup founders should solicit ideas, discuss them and express gratitude for them. Give your employees time, and demonstrate an interest in knowing what they are doing periodically.

The key for Kaizen success at an individual level involves starting and developing internal conversations within the mind. It also involves explicitly jotting down and tracking your progress, while pursuing small, non-threatening and actionable steps in a gradual manner. This can include visualizing oneself in the state of results you want to accomplish for the startup. It can be repeating positive and powerful affirmations or statements to yourself just to reinforce the idea within your subconscious mind similar to persuasive advertising that seeks to hammer a point into your subconscious until your actions are guided by the firmly embedded thought.

This can begin with devoting a minute a day towards the pursuit of focusing on the tiniest things along even if they seem absolutely ridiculous. It makes the change pleasant, effective and simple in the long haul. Even during crisis situations, the philosophy of Kaizen helps in breaking down large and seemingly insurmountable tasks into baby steps.

Chapter Five: Idea-Sharing and Kaizen Boards

One of the most important aspects of Kaizen is idea-sharing to make every employee feel a valued part of the organization, which makes creating an idea-sharing important for any startup or small business. How do you manage and document these ideas before they are lost? How can these ideas be used for reference? What are the different idea management systems that can be used to leverage the power of your employees' ideas? Here are some ideas for managing your startup's idea-sharing by using the power of Kaizen boards.

A Kaizen board is nothing but a visual tool that keeps track of every idea for improvement and enhancement of processes gathered throughout the organization and later helps analyze the status of these ideas by tracking them through improvement processes. Every employee is encouraged to be a part of the improvement process to give the business an edge over competitors. As a startup, you can visualize ongoing and planned improvements for giving yourself a clear way to track progress through the implementation of ideas. This way, you know which ideas work as well those that do not.

I know some organizations where they have a dedicated idea-sharing day every week, where employees are encouraged to share their ideas for improvement and enhancement. These ideas are recorded on the Kaizen board and immediately implemented within the next few days. People are asked to record their progress post the idea implementation on the board to witness real, tangible results of the idea-sharing process. This also motivates employees to know that they are a part of the idea-sharing process and makes them contribute even more ideas and feel a valued part of the organization.

Think about a hectic day where you are having a supremely productive meeting. This is a team meeting where everyone has shared some amazing ideas and insights about working smart and making the process more efficient. You know you can increase productivity and get a lot more done by implementing these ideas. Now, a couple of weeks pass and none of the ideas or suggestions are implemented. There is another weekly meeting to discuss ideas. Team members come together again to enthusiastically share even more ideas. All that's done is sharing ideas without actually implementing these valuable ideas because somewhere along the way, these ideas are lost.

This is exactly where a Kaizen board can step in and fill the gap. I know a lot of startups and small businesses think this to be a lot of work (keeping a dedicated board for ideas) though it doesn't seem like any real work. However, an idea board can get the entire team together to turn in great ideas into actionable affairs, which improve the group, team of business' overall deliverability. As long as the entire organization is committed to reviewing the Kaizen board as a team, and gives it priority over other work, a Kaizen board is capable of helping you transform ideas and vision into reality.

Kaizen boards come in a variety of sizes and forms. One way to come up with suggestions for improvement or quality control of processes is to list the problems your startup is facing on a large sheet of paper. It can be anything from high marketing or promotional costs to low customer satisfaction to increased manufacturing defects–just about anything. Allow every employee to offer solutions to these problems by brainstorming. The "before and after" image forms can work brilliantly for problem-solution brainstorming.

One of the biggest advantages of Kaizen boards is that they are public and accessible to every employee, not just a few key decision makers. This way, every employee feels a part of the decision making process. You may not have considered a problem where a competitor has a clear edge over you. For example, they may have created a solid social media promotion strategy that may have led to reduced marketing and promotional costs. A social media savvy employee may have noticed this and suggested that to keep your marketing and promotional costs low, as a startup you should build a strong, loyal social media community through engagement and interaction with customers. When this idea is accessible to everyone, it becomes easier to implement it. Keep your Kaizen board transparent, open and accessible. Unlike some companies which are secretive about their improvement and change implementations, ensure that everyone has access to these ideas to facilitate quicker, easier and a more seamless implementation.

With Kaizen boards, as a startup, you can bring out the creative talents of employees with your organization. Generally, when there are problems or defects, we are quick to blame other people for it. At times, we'll pretend like it didn't happen at all. This goes against the Kaizen approach. Each of these mistakes must be brought out into the open so that they turn into learning experiences. There are plenty of hidden opportunities for improvement in these mistakes, and hiding them only deprives your team of an opportunity to learn and improve.

Whenever you are an employee makes a mistake, ensure that you accept it and share it with everyone as a learning. Employees should be encouraged to share their mistakes with each other as a tool for facilitating continuous changes and improvements to drive home the Kaizen philosophy. Even when you as a key decision maker makes a mistake, call a meeting and discuss it in detail with your employees.

This is one of the most effective ways to bring about change and improvement, which can offer you a competitive edge over other similar businesses. Process improvement or enhancement happens only when mistakes are brought to light, and ideas/suggestions are offered to enhance these processes. This optimizes errors and boosts productivity to give you a clear competitive edge over similar businesses.

Even if something is working well for you, someone who is closely working on the process and systems can come up with a better or more efficient way to do it. This helps optimize your productivity, profits and results to give you an edge over competitors who may be stuck with the same old ways of doing things.

Startups are a wonderful platform for practicing Kaizen because, by their nature, there are no set systems or processes in place. They are innovative, breakthrough and cutting-edge, which means you can employ plenty of new ways to do things which gives you a clear edge over other established businesses that have been doing the same thing for several years. You have the opportunity of inventing new systems and processes to bring about breathtaking results. Kaizen boards are powerful tools to help to keep track of your organization's progress in an easy to use and highly visual manner. It is a record of your current and future Kaizen efforts to help you stay on track with the suggested changes.

Understand that more than anything else, Kaizen is a mentality. It isn't as much a method as it is a way of life. It is the way you approach your work through continuous training, analysis, experimentation and data-channelized decision making. If you are practicing continuous and consistent improvement, Kaizen is firmly embedded in all aspects of life.

The Job Methods Improvement Model

This is another popular idea documenting the process that involves breaking down each job or task into its basic components. Once that's done, the task is improved by removing, combining, reorganizing and simplifying the process. Later, change management is practiced by proposing these ideas to the boss. Finally, the improvement impact is measured through a set of predefined criteria such as quality, safety, quantity, and cost. The JMT method is about identifying new ideas and insights, and different ways of practicing things.

As a startup, look for small ideas that can be improved upon gradually instead of attempting to give an entire department or business a makeover because something isn't working in your favor. You can't plan an entire department layout in a day. However, small changes can be implemented on a daily basis. Don't run after a big installation or new equipment fitting. A startup doesn't require major changes or improvements. There are hardly any systems in place currently. This is the time to set big systems and make small changes.

This idea model also involves the "train the trainer" method used for growing talent, skills and improvement by spreading a methodology widely. Post-war, the Japanese lost much of their skilled laborers and has to manufacture high-quality goods in massive quantities to make for the shortfall of manpower and growing demand. They had to produce a high quantity of goods at competitive rates to resurrect their economy. This could only be possible by finding more efficient, cost-effective, productive and easier way to do things. They had to optimize their limited resources to produce maximum results.

Vision Boards

Vision boards can also drive the Kaizen philosophy home

effectively. They are boards that graphically depict where you want to reach as an organization. You can place a large board in the center of your office to facilitate easy and open access to the board. Team members or employees can keep accessing these boards to stick images of where they want the organization to be. This aspiration-based vision board can have another visual board alongside it to reveal real results accomplished as a result of the aspiration-based improvements.

For example, let us say you want to increase your social media reach a million followers to curb your marketing budget. You mention that on the aspiration based vision board. You graphically or pictorially depict having a million followers for your business page or business niche driven community. Then everyone works towards building a social media following for the page, which in turn reduces your organization's marketing and promotional costs. You track your progress on the results visual board to include details such as how many followers your business or community page has managed to gain, and how it has helped in reducing your marketing and promotional budget over a period of time.

Conclusion

Thank you for making it through to the end of *Kaizen for Small Business Startup: How to Gain and Maintain a Competitive Edge by Applying the Kaizen Mindset to Your Startup Business and Management-Improve Performance, Communication & Productivity,* let's hope it was informative and able to provide you with all of the tools you need to achieve your goals.

Like everything we take up, there will be a learning curve in building and running a successful startup. However, when you create the right environment and culture within the organization, the process can be more rewarding and fulfilling—nothing that's worth having ever comes easily.

The Japanese Kaizen philosophy has been successfully implemented by thousands of big and small businesses to drive massive results and build an atmosphere, improvement, quality control, as well as employee and customer satisfaction.

Practice these Kaizen management hacks, and make the most of the edge that you are just starting out. The rewards of making small changes and improvements to your business can be tremendous—even though these changes may not be *immediately* visible.

If you truly enjoyed reading this book, please share your thoughts by writing a review on Amazon. It would be greatly appreciated.